beekeeping
adrian and claire waring

For over 60 years, more than
50 million people have learnt over
750 subjects the **teach yourself**
way, with impressive results.

be where you want to be
with **teach yourself**

Illustrations by Barking Dog Art.
Photographs © Adrain and Claire Waring.

2 3 4 5 0 5 2 2 /

For UK order enquiries: please contact Bookpoint Ltd, 130 Milton Park, Abingdon, Oxon, OX14 4SB. Telephone: +44 (0) 1235 827720. Fax: +44 (0) 1235 400454. Lines are open 09.00–17.00, Monday to Saturday, with a 24-hour message answering service. Details about our titles and how to order are available at www.teachyourself.co.uk

For USA order enquiries: please contact McGraw-Hill Customer Services, PO Box 545, Blacklick, OH 43004-0545, USA. Telephone: 1-800-722-4726. Fax: 1-614-755-5645.

For Canada order enquiries: please contact McGraw-Hill Ryerson Ltd, 300 Water St, Whitby, Ontario, L1N 9B6, Canada. Telephone: 905 430 5000. Fax: 905 430 5020.

Long renowned as the authoritative source for self-guided learning – with more than 50 million copies sold worldwide – the **teach yourself** series includes over 500 titles in the fields of languages, crafts, hobbies, business, computing and education.

British Library Cataloguing in Publication Data: a catalogue record for this title is available from the British Library.

Library of Congress Catalog Card Number: on file.

First published in UK 2006 by Hodder Education, 338 Euston Road, London, NW1 3BH.

First published in US 2006 by The McGraw-Hill Companies, Inc.

This edition published 2006.

The **teach yourself** name is a registered trade mark of Hodder Headline.

Copyright © 2006 Adrian and Claire Waring

The publisher has used its best endeavours to ensure that the URLs for external websites referred to in this book are correct and active at the time of going to press. However, the publisher and the author have no responsibility for the websites and can make no guarantee that a site will remain live or that the content will remain relevant, decent or appropriate.

Typeset by Transet Limited, Coventry, England.
Printed in Great Britain for Hodder Education, a division of Hodder Headline, 338 Euston Road, London, NW1 3BH, by Cox & Wyman Ltd, Reading, Berkshire.

Hodder Headline's policy is to use papers that are natural, renewable and recyclable products and made from wood grown in sustainable forests. The logging and manufacturing processes are expected to conform to the environmental regulations of the country of origin.

Impression number   10 9 8 7 6 5 4 3 2 1
Year   2010 2009 2008 2007 2006

contents

introduction

In writing this book, we have tried to produce something that will give the potential beekeeper a useful introduction to the craft. Beekeeping is the kind of hobby where many aspects, such as entering honey shows, can be expanded upon and become a hobby in their own right. Some beekeepers develop their craft from a pastime into a full-time occupation. Think of this book as something to whet your appetite. We began in this way many years ago and are still interested and always finding new facets about which there is more to learn.

Read this, and any other books you choose, carefully. The most important information you need to learn is how a colony works and its various stages of development, particularly in relation to your local environment.

Between us, we have kept bees for 65 years, so far. We have had direct experience of beekeeping in Scotland and the north and south Midlands of England. Bees vary according to their racial type and their adaptation to the local climate. However, what bees 'do' and when they do it is brought forward or held back by variations in temperature. What this book says can be helpful wherever bees are kept. If you, the beekeeper, use your common sense.

Our aim, in this book, isn't to tell you everything there is to know about beekeeping, but to tell you many of the things *we* would like to have been told when we began. You will be missing a great deal if you try to work in comparative isolation. You can gain so much from others and we would encourage you to seek out fellow beekeepers, particularly within your local beekeeping association.

What this book cannot give you is experience, but we hope it will help you enough so that the experiences you have will be meaningful and not incomprehensible. Try to look for the common denominator in all explanations and you will see the similarities.

We hope this will be the first book you acquire in your bee book collection (another extension hobby!). We now have over 700 books dealing with virtually all aspects of bees, beekeeping and associated subjects from 1658 to the present.

Since the oldest bee book in our collection was written, European bees have been taken to all parts of the world, including North America, where such bees were previously unknown. What we say about the bees themselves will therefore be true wherever they are found. The equipment may be different or varied but bees are much more constant. Whatever the type of hive or equipment, beekeeping methods will be modified by local honey flows and weather conditions.

Wherever or whatever else you do in beekeeping, enjoy it. It is not as dangerous or as expensive as, say, riding a motorbike. It has some of the same qualities as the hobby of fishing without the need, as it seems to us, to sit around in cold, wet places! This book will not reveal all there is to know about beekeeping but we hope that it will be a good beginning.

Adrian and Claire Waring

01

**naming
the parts**

In this chapter you will learn:
- about the honey bee from
  head to toe
- how the colony operates
- about honey bee food.

So, you want to start beekeeping? There are many ways to do this, but all the various methods need you, the prospective beekeeper, to understand firstly what you will be dealing with.

Bees are insects. They have the typical insect structure. By this, I mean a hard outside, the exoskeleton, which is divided into three parts – the head, the thorax and the abdomen. There is a narrow neck between the head and thorax and an equally narrow waist, the petiole, between the thorax and abdomen. These two connective parts allow the insect some degree of flexibility.

Senses: smell, touch and taste

The head has important sensory organs. The first of these are the antennae, commonly known as 'feelers'. They do have important senses of touch but they are the equivalent of the bee's 'nose' as well. They are capable of detecting scents important to bees at concentrations thousands of times less than the level that would permit them to be detected by humans. There are also organs of taste on the antennae, as well as on the tongue and forelegs. The head also carries the mandibles, or mouthparts, and the eyes.

Vision

Bees have three 'simple' eyes on top of their heads. These are thought to function as detectors of light levels. Inside the head are salivary glands which produce moisture, and brood food glands which produce the substance which is fed to larvae.

Bees have two compound eyes, one on each side of the head. Essentially these are patches of smaller eyes, each with its own nerve leading to the brain. The compound eyes can see ultra violet as a colour and are also aware of polarized light. This means that if the bee can only see blue sky, it 'knows' where the sun is.

Each eyelet of the compound eye is made of several cells which are each sensitive to light vibrating in a separate plane. The earth's atmosphere has a similar effect on light to that of polarized sunglasses. Because of this, the bee, even if it can only see a little patch of blue sky, will know where the sun is and can navigate using this.

Walking and flying

The next part of the insect body, behind the head, is the thorax. It is simplest to think of it as the bee's chest. All the body parts that give the bee mobility either on the ground or in the air – legs and wings – are attached to the thorax. It is the site of the main muscle pack in the bodies of all insects.

Bees have four wings, the pair on each side hooking together in flight. Muscles attached directly to the wing extend them and are capable of altering the wings' angle. The direct flight muscles turn the leading edge of the bee's 'combination' wing so that the insect achieves lift in both the upstroke and the downstroke of the wings. This rolling action also enables the bee to hover very effectively. The wings are moved up and down by muscles that are not attached to them at all. They work by distorting the shape of the thorax very rapidly indeed. By this means, the wings beat at 250 hertz (beats per second) and this produces the 'hum' we hear.

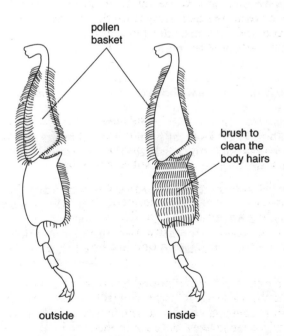

pollen
basket

brush to
clean the
body hairs

outside   inside

figure 1 hind legs of the honey bee

The drone is the male bee which does not sting. If you hold him gently between thumb and forefinger, placed top and bottom of his thorax, allowing the wings to beat freely, you will feel these rapid distortions as a vibration against your fingers.

The bee has six legs with six main joints in each. Each of the first pair of legs carries a device for cleaning the antenna. Each of the hind pair carries a device for carrying pollen (see Plate 1), and all legs have brushes on their inner surface to brush the body hairs free of pollen. Each leg ends in a foot which comprises two claws which can hook onto surface roughness. If these claws fail to grip, a central pad comes into play and this allows the bee to find a grip even on smooth surfaces, such as glass.

The abdomen

The abdomen contains the bee's heart, honey stomach (or crop), true stomach and intestines. In the honey bee female, there are the ovaries and the sting; the male's abdomen contains his reproductive organs, the testes, but he has no sting. The crop holds nectar and pollen. In foraging bees, the nectar is regurgitated upon return to the hive and passed to house bees which store it in honey cells. The pollen passes on into the true stomach. In theory, the crop can hold a load of 100 mg, but most weigh only around 40 mg. Ovaries are found in the two female forms, the queen and worker, and their function is to produce eggs. Generally, this only happens in the queen but there are situations where the worker's ovaries can also function.

The bee's intestines work in a very similar way to our own. Food is digested in the true stomach and the nutritious part is absorbed by the intestine. The intestine ends in a large chamber where waste is stored for a while before being excreted. In healthy bees, this act takes place when they are in flight. Some diseases rely on making the bee defecate on its comb as a way of ensuring its transfer to other bees. This large chamber is known as the rectum and it is capable of expanding to fill a large part of the abdomen. In the winter, when it is too cold for bees to fly from the hive, it can apparently fill the abdomen. In the UK, especially in the north, a bee may be required to hold the waste products of six or seven weeks of food consumption and digestion. Beekeepers traditionally know the early spring flights after a period of long confinement as 'cleansing flights'.

The ancestors of bees and wasps are sawflies. These use the barbs on their stings to cut into plant tissue to provide places in which they lay their eggs. In the modern bee, the two barbed shafts of the sting mechanism slide upon the third unbarbed part. The two move alternately so that one side moves forward, penetrating the tissue of whatever has been stung. When the other one moves, the barbs on the first shaft stick fast so that the second one digs in further. The sting mechanism has its own set of muscles and nerves and can operate for a short while on its own, when detached from the bee. Once inserted into elastic human skin, the sting can continue to dig its way in and the muscles can inject the full amount of venom (see Plate 2). The best way to reduce the effect of this venom is to remove the sting as quickly as possible in whichever way is easiest at the time. Try to avoid squeezing the sting; scraping is best.

Bees use stings for the defence of the colony. The venom produced causes pain and swelling. Within a week or two of starting beekeeping, most beekeeper's dogs know the difference between a bee and a fly and leave bees strictly alone – which is what the bee intended.

When stung, most people feel pain. The area where the sting went in will swell. This swelling can be spectacular and off-putting. However, if the reactions are confined to the area stung, then all is fairly normal. Taking antihistamine tablets can help. If you know you are going to inspect your bees, take the antihistamines before you go to the apiary to allow them to get into your system. However, if reactions occur all over the body like a nettle rash, or you experience more severe symptoms, consult a doctor at once. A small proportion of people react very badly indeed to bee stings and can be in danger of dying from anaphylactic shock. This is an extreme allergic reaction. A simple swelling where you were stung is a perfectly normal reaction. As the swelling goes down, it is often accompanied by itching. The more you are stung, the less your reaction will be. With me, the itching went very quickly, followed by the tendency to swell. The face is particularly susceptible to swelling, but modern bee veils are so good that stings to the face occur only if it is uncovered. Good protective equipment is essential and its proper use reduces accidents to a minimum.

If reading this has put you off beekeeping, then all well and good. It is not a hobby for everybody and not to be undertaken if you are not wholeheartedly interested.

Colony activities

The honey bee, (*Apis mellifera*), lives in colonies. Honey bees fly alone to collect food for the colony. However, individual honey bees, separated from their colony, do not live for many days unless they are accepted by another colony.

Bees breathe through openings in their bodies called spiracles. Air is pulled in and passes through tubes, or trachea, partly by its own pressure and partly as a result of the bee's body movements. These trachea branch and get smaller and smaller down to the astonishing level of providing oxygen to individual cells. The prime example of this is oxygen provision to cells in the flight muscles in the thorax.

This method of breathing limits the body size of insects. Social insects, such as honey bees, form a colony which functions in ways that allow us to compare it with the single body of a mammal. This sacrifice of individuality, known as altruism, allows the social insect to gain some of the advantages of being big because the colony as a whole acts in a way similar to a large animal.

Comb and comb building

An integral part of colony life is the combs that the bees build. The combs are used to rear young bees and store food. These are made from beeswax, secreted by the workers, and are mostly composed of hexagonal cells. The combs are double-sided, with the cell bases meeting at the midrib. In natural comb built freely by a colony in an empty cavity, the distance between these midribs varies only a little, a space of roughly 36–8 mm (see Plate 3).

When building comb, bees hang from the upper surface of the cavity in which they have chosen to live. Small plates of wax are formed by the wax glands on the underside of the abdomen. These are then brought forward to the bee's head and chewed with the mandibles before being stuck to the surface of the cavity. Eventually, lines of wax are deposited. Bees often make use of an edge or an imperfection in the surface for the beginning of the comb and the rest develops around this. Initially, the bees extend their line downwards and build a midrib. As this grows, the cell bases are formed on both sides. Cell bases on opposite sides are offset so that the point of contact of three cells on one side is in the centre of the one on the other. The leading edge of the midrib advances as bees continue to build. Behind the leading edge, the cell walls are extended. In a free space, all this work is carried out within a cluster of many long chains of bees formed by the individual insects locking their legs together.

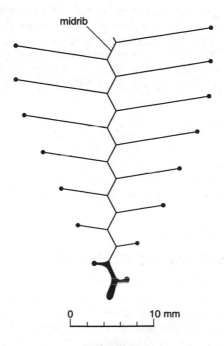

figure 2 side-view cross-section of honey bee comb under construction – note the cells are off-set either side of the midrib

Cells are built with two parallel sides orientated vertically pointing to the surface on which the comb is first attached. At first sight, the cells in freely built comb seem very accurate, but close examination will reveal that bees are like most home DIY enthusiasts – they make things fit. Comb can be started in several places within the cavity and then joined together in due course. Sometimes you can't see the join, sometimes you can.

Bees are quite capable of attaching their comb to a vertical rather than a horizontal surface. The cells start off with the 'vertical' sides pointing to the surface. As the comb is built, gravity takes over and it describes a graceful 90-degree turn and the natural comb then has the two parallel sides horizontal instead of vertical. Bees can also build comb upwards. They usually do this when they are given, or take over, a space containing no frames or foundation. Such combs are often curved and I have always supposed that this was to give the structure some strength since new comb is quite soft. There is also a height limit on such comb and it never seems to exceed 10–12 cm. If the bees store honey in such comb, its weight will hasten the collapse of the comb. Although beeswax becomes softer and more pliable as the temperature rises, inside a bee colony it has all the tensile strength required. Once comb has been used to rear brood, it is made much stronger by the cocoons left behind by the emerging bees.

Initially, an average size colony in a new cavity will build five to six combs and extend them as required. Under the influence of queen substance, the bees will build only worker cells. In the following season, they will build a proportion as drone cells. These are much larger, just over four per 25 mm, to accommodate the larger male bee. Between areas of the two types, worker and drone, you may find cells of very odd shapes which are just used for storage of nectar. These are called interstitial cells. Bees will also brace one comb on another with horizontal 'bars' of wax, often constructed with a recognizable cell shape or even with complete cells. Some bees build a lot of this brace comb which can make it difficult to handle the combs.

Bees and their pheromones

Honey bees store their energy food, nectar, as honey. This is a form which allows it to keep through the flowerless days of winter. The mammal lays down a store of fat for the same reason. Mammals' blood carries food to tissues. It also transports chemical messages which cause actions required by the body. Bees constantly share food which contains chemical messages (pheromones) and this circulation obtains similar results in the colony. A pheromone is defined as a chemical substance produced by an individual which affects the behaviour or physiology of other members of the same species. A sting, for example, is accompanied by an odour (isopentyl acetate) which activates other bees to come and sting in the same place. Much of the order within the colony is governed by the presence or absence of pheromones, particularly those produced by the queen, and these are passed around the colony by contact and this food sharing/circulatory system just as our hormones are transported via our blood system.

Female castes

Honey bees have two sexes like most living things and one of these, the female, appears in two forms, or castes: the queen and the worker (see Plates 4 and 5). Any fertilized egg laid by a queen honey bee can develop into either of these two forms. The determining factor is how the larva is fed and on what. A worker larva can be raised as a queen rather than a worker if it is fed copious amounts of royal jelly throughout its larval life.

The queen honey bee is larger than the worker and longer than both the worker and the male drone (see Plate 6). However, she is not as bulky as the drone. Her head is superficially very similar to that of the worker but she has fewer individual eyes in her compound eyes – estimates or counts vary from 3000 to 6000 per eye. The main cause of this variation is probably human error. I am more inclined to believe the lower numbers.

The queen, like the worker or drone, has jaws known as mandibles which move from side to side. They look like two little stiff arms that meet in the middle of the mouth. They can grip small objects and even be used to bite. Inside the queen's head are glands which discharge their product onto these mandibles. These mandibular glands are of vital importance to the honey bee colony. Their complex pheromone secretion (queen substance) causes a number of things to happen. Its presence makes worker honey bees rear female eggs as more workers. If there is no queen substance at all in the colony, some young larvae are reared as queens. Queen substance also represses the development of the small ovaries which are part of the worker's physiology. Prolonged absence of queen substance allows these ovaries to develop and some of the workers start laying eggs. These are often placed two or three to a cell and I have seen as many as six together. I have also seen two or three larvae in a single cell. These always develop as drones. If the workers lay eggs in drone cells, the resulting adults are full-sized drones, but if the eggs are in worker cells, the emerging drones are small. Whatever size they are, they are capable of mating. A colony that has reached this situation is doomed to die so you can regard this apparently pointless activity as an attempt by that particular colony to keep its genetics within the bee population.

When the colony swarms, queen substance keeps the bees together in flight and stabilizes the swarm cluster. If you remove a queen from a colony, bees will start looking for her within 20 minutes. Within five or six hours, they will have started to produce replacement queens from young worker larvae.

The queen's abdomen is long and tapered. It contains two bundles of tubes which produce the many eggs she is capable of laying. Each bundle is known as an ovary and each tube (ovariole) can produce eggs. There may be over 150 ovarioles in each ovary. The queen is fed on demand by workers and, at her egg-laying peak, is reputed to be able to lay more than her own weight in eggs per day, suggesting that she eats more than she weighs per day. The queen is a slim insect, 2–3 cm long. The more eggs the queen is laying, the longer she is. Most queens in full lay are too 'fat' to fly.

A queen's life starts when she is the single unmated queen in a colony, known to beekeepers as a virgin queen. After a colony has produced such a queen, she has to fly outside of the hive to mate. These mating flights take place during the first 10–14 days of her life. She will make a number of exploratory learning flights and on one of these, usually the last, she will mate on the wing with 10–20 drones. The sperm she receives migrate to a storage sac, the spermatheca, inside her abdomen. A few sperm from here can be deposited on an egg as it is laid. This happens to every 'female' egg but not to those destined to produce males.

A laying queen walks over the surface of the comb. She tends to lay in a pattern, working outwards from an initial central point and following a spiral path. Having laid in an area on one side of the comb, she will lay a similarly sized patch on the opposing face of the next comb in the same relative position. She lays in cells that are warmed to broodnest temperature and that have been prepared by the worker bees. You can see evidence of this, when you look at comb in daylight. The cell bases appear polished. In the darkness of the broodnest, the queen cannot see this but she can probably smell what the workers have been doing. Having laid eggs on the next side of the comb, the queen will return to the first side to lay more eggs, and so on. Obviously, this is not invariable but it is the pattern of egg-laying that you will learn to recognize. Since development of worker brood takes three weeks, the queen can return when the first batch of eggs has emerged as adults and lay again in the same cells. This results in a pattern of concentric rings of brood of similar age: beekeepers know this as the brood pattern.

Borrowed on 02/11/2013 16:05 Till

1) Haynes bee manual :the complete
   step-by-step guide to keeping bees
   Due date: 30/11/2013
   No.: 181961821
2) Teach yourself beekeeping
   Due date: 30/11/2013
   No.: 234505221
3) Keeping bees :a complete practi
   guide
   Due date: 30/11/2013
   No.: 237928821

Total on loan : 3

2/11/2013 - 16:05

Items on loan on 01/10/2013 16:32 Ti

1) Haynes bee manual :the complete
   step-by-step guide to keeping bees
   Due date: 29/10/2013
   No.: 181961821
2) Teach yourself beekeeping
   Due date: 29/10/2013
   No.: 234505221
3) Keeping bees :a complete practica
   guide
   Due date: 29/10/2013
   No.: 237928821

Total on loan : 3

01/10/2013 - 16:32

If there are only a few cells on a full brood comb that are empty, this pattern shows you that the eggs, larvae and pupae have a high rate of survival – a good brood pattern.

Before laying, the queen puts her head into the cell, her antennae pointing forwards. She braces her forelegs across the diameter of the cell. If she is satisfied that the cell is properly prepared, she comes out of the cell and walks forward slightly so that she can insert her abdomen into it. The egg that she lays stands upright on the cell base. Many think it slowly topples over before it hatches but I am not convinced that this is the case.

It has been found that when the queen inspects each cell, the diameter that she measures with her forelegs tells her what type of cell it is. The majority of cells are designed for rearing worker bees. If the queen inspects a worker cell, she will deposit sperm on the egg just before it is laid to fertilize it. If it is a wider drone cell, she does not deposit sperm on the egg and it remains unfertilized. However, such eggs are fertile because they hatch, and the larvae develop and eventually emerge as drones. Thus all female eggs contain the genetic material from both drone and queen whilst the drone carries only half of those genes – those from his mother. You could say he has a grandfather (the queen's father) but he does not have a father!

Queens are raised in special cells constructed during the summer (see Plate 7). Workers are reared in hexagonal worker cells with about five and a half cells every 2.5 cm. Drone cells are similar but larger at just over four per 25 mm (see Plate 8).

Queens can live for several years – three years is quite possible. I once owned a queen that reached the advanced age of five years, but few beekeepers allow their queens to live to anything like that age. The best advice I can give is to quote Charles Mraz, the American queen breeder, 'Never kill a good queen.'

Drones

Drones are the males and have the reputation of being lazy (see Plate 6). Shakespeare's 'lazy, yawning drone' or PG Woodhouse's 'Drone Club' are often the ways that members of the general public think of drones. However, they are vital to bees as a species. They carry half of the bee chromosomes and behave in a way that reduces inbreeding to a minimum. In the United Kingdom, they fly from colonies only between about 10 a.m.

and 5 p.m. They fly to drone congregation areas where drones from many colonies assemble. Virgin queens also fly to these areas and this is where mating takes place. The area has an edge to it and within the area, drones investigate anything in the air. Even stones thrown up through the air are temporarily followed. Queens go to these areas to mate with up to 20 drones.

The drone has large compound eyes each containing about 9,000–13,000 individual eyes. His eyes are so large that they meet at the top of his head, forcing his simple eyes down onto his forehead. His antennae are larger than those of the queen and worker and they carry many more sensory cells. His tongue is short and his mandibles are toothed, like those of the queen, but they are small, uncoordinated and weak. Drones are mainly fed by workers but they are capable of feeding themselves. However, because of the drone's short tongue, the cells have to be nearly full for him to be able to reach the honey. Drones are there to mate with virgin queens. Finding their own food would be a waste of time and energy.

The drone's flight is very rapid and strong. He appears to be able to fly faster than either the queen or worker. Drones live for about eight weeks. They are sexually active at around two weeks of age and remain so for about three weeks. They are first produced when the colony starts getting strong in the spring and cease to be produced by late summer. Workers either throw them bodily out of the hive or simply neglect them to death.

Workers

The worker honey bee works (see Plate 9). The worker is female, one of the two castes. She has smooth, spoon-shaped mandibles that are employed in many tasks which tend to be age-related. Newly emerged workers do nothing at first and then follow a succession of tasks. For example, they feed young larvae at first, when their brood food glands are working well. Older larvae are fed when the glands reach a later stage of development.

Many years ago in Germany, a bee was observed when it emerged in an observation hive and a record was made of its actual activities. The greatest total length of time was spent resting. The next big chunk was spent patrolling or looking for work and the smallest part was spent actually working. Such a lifestyle may sound very familiar, especially to those reading this in the depths of their armchairs! However, in a honey bee colony, a great deal gets done. Bees don't work on a task and finish it. They work for a while and then move on. Another patrolling bee takes up that task, and so on. They are probably guided to the job and prompted to work by pheromones and maybe the suitability of the job for the capabilities that they have because of their particular age. Therefore, in one day, a bee will rest between tasks such as feeding a larva, trimming cells, capping honey cells or capping worker brood cells. As the workers age, their wax glands on the underside of their abdomen develop and they make beeswax when required. Finally, at about three weeks old, their life half gone, they start to fly outside to find pollen and nectar.

A small percentage of the bees in a colony specialize in removing the dead. Within a very short time of death, changes in the composition of the corpse trigger a response and the dead bee is dragged to the entrance and dropped out. Very often, the undertaker gets a good grip and flies off with the body. I have seen them drop corpses up to ten metres away or even fly out of sight. However, this behaviour is only seen when the weather is relatively warm. On cold days, the bodies are simply dropped from the entrance.

There is another small group of bees within the colony which scout for new sources of food and return to tell others in the hive. A smaller number still, specialize as guards, inspecting all bees that try to enter the hive, particularly when no nectar is being carried. Alarm pheromones are released if the hive is invaded and can result in an overwhelming response as other workers join the guards.

Winter bees

It could be said that worker bees can be found in two forms. 'Summer' bees are essentially as I have described them. Workers produced from around August onwards, however, can be slightly different internally from those produced during the summer. They eat a lot of pollen and, as a result, clusters of fat cells are featured in their abdomens. Their brood food glands remain developed in the same way as those of a young bee and they are able to tolerate long periods of confinement in the hive during winter. Their lives are greatly extended from the usual six weeks in summer to several months, possibly up to six, over the winter. To beekeepers, they are known as 'winter' bees. It is interesting to think that workers emerging in October may well be collecting nectar in the following spring.

Nectar and pollen

Nectar is a solution of various sugars. This is converted into honey which forms the energy-rich carbohydrate part of the bee's diet. Nectar is produced by plants as an insect attractant because many employ insects to move pollen, the male element of plant sex, to the female part of another flower of the same species. Pollination mechanisms are devised by the plant to attract insects and make sure that its seeds are fertilized. However, to the insect, plant pollination is merely an accidental by-product of its food-collecting activities.

Plants produce far more pollen than they need because it is used by many insects as food. The bee does not know that its activities result in pollination, it is merely collecting its food and is also intent on collecting enough to be able to store a surplus as winter stores. Water is evaporated from nectar to produce honey, which will keep over a period. We humans collect that surplus. In the United Kingdom, our honey bees produce something like 4,000 tonnes of honey and we eat a further 20,000 tonnes which is imported. However, the value of crops produced by pollination has been estimated recently for the United Kingdom at around £130 million per annum. It must be said, however, that honey bees are not the only pollinating insects involved.

It is believed that, originally, honey bees developed in a tropical environment. They probably had a wasp-like ancestor and evolved to obtain the protein they required from pollen. The need to collect and store food probably helped these ancestral insects to survive hot dry periods when there were no flowers. The bee species found in the tropics today do exactly this.

The colony and winter time

As the last ice age came to its gradual end and the various bee species extended onto the land exposed by the retreating ice sheet, this storage activity also helped bees to survive the flowerless periods. The natural northerly limit for places where honey bees can live is considered to be where rivers remain unfrozen for seven months of the year.

The instinct to store food, therefore, in both the tropics and the temperate regions, allows colonies to survive so that honey bees are present in strength, ready to take full advantage of what early-flowering plants have to offer. Bumble-bees and wasps overwinter as mated queens which have to start from scratch each spring to establish and build up their colony. Some plants, such as ivy, hellebores and willow, flower in the late autumn, winter and spring to get the full attention of honey bee foragers. This 'coming together' is mutually advantageous to both bees and plants.

As a colony, bees can maintain their own temperature. The individual bee is cold-blooded and becomes motionless at 5 °C and it then slowly dies. At 14 °C, bees in a colony begin to cluster together and use their bodies as insulation. If the colony is rearing eggs, larvae and pupae, at an outside temperature of around minus 10 °C, the edge of the cluster will be at 6–7 °C and the centre at 35 °C. This winter cluster can be very efficient. Most colonies rear some brood during the winter. This may be continuous or intermittent. It may be in comparatively large amounts or not at all. Bees with a brood nest within the cluster at 35 °C have been known to maintain this when the outside temperature falls to minus 35 °C. The fuel for all this is, of course, honey.

Bees living in large colonies with large amounts of stores, often provided by the beekeeper, look fairly invulnerable. However, surveys were carried out on wild colonies living in deciduous woodland in America, before the arrival of the varroa mite (see Chapter 8) which has now essentially wiped them all out. These showed that 75 per cent of wild colonies died during their first year or their first winter. The main cause of colony demise was starvation.

02

the honey bee year

In this chapter you will learn:
- how colonies survive the Wintertime
- about spring build-up
- about swarming time.

Beekeepers exploit bees. Beekeepers consequently have a duty of care to their bees. They feed them to supplement their honey stores, and help them to withstand their natural enemies and cure them from diseases. If it was not for beekeepers protecting bees from the exotic mite pest *Varroa destructor* (see Chapter 8), it is doubtful that more than one in a thousand colonies from our basic population would be left alive.

Autumn and winter

If it is to have any chance of survival, a natural colony of bees will be living in a cavity. Bees in the United Kingdom very rarely choose a home in the ground and tend to prefer a hollow tree or similar cavity. Relatively recent experiments have shown a cavity of about 40 litres is the optimum size chosen by a colony. The colony builds vertical, parallel beeswax combs, suspended from the top of the cavity. The object is for the colony to produce enough bees to be a viable unit of 10,000–20,000 to live through the winter and to have honey stores in the region of 15–20 kg.

Strong colonies winter the best, providing they are healthy. To a beekeeper, a strong colony means one with lots of bees, hence the beekeeping maxim in the United Kingdom, 'The best packing for bees – is bees.'

Where the colony is rearing brood (eggs, larvae and pupae), it maintains a temperature of around 35 °C at its centre. This brood nest is orientated on the entrance. In the autumn, the brood nest will be producing only worker bees. Drone production stops in late summer. The queen's daily egg-laying rate get less and less and, by winter time, is likely to cease altogether. The brood nest is roughly spherical and around it is a pollen 'shell'. It is rather like one of those Chinese puzzles where one ball is carved inside another. The pollen band is wider above and behind the brood nest, narrower on the side facing the entrance to the colony, and narrowest, if it exists at all, below the brood nest. Imagine the parallel combs as if they are slices of this sphere. Irrespective of the way they pass (front to back or side to side), the arrangement of the brood and pollen in each comb is the same, relative to the entrance. This is just the same as if you slice an orange in two directions at right angles to each other.

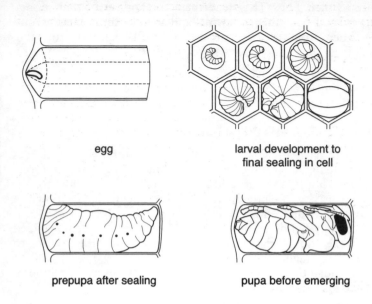

egg

larval development to
final sealing in cell

prepupa after sealing

pupa before emerging

figure 3 the honey bee life cycle

The cluster

As the weather becomes cooler, bees start to clump together until, by the time the ambient temperature has reached 15 °C, they have formed a cluster. As the temperature drops further, bees squeeze together and the cluster becomes smaller (see Plate 10). The bees are mostly the longer-lived winter bees. In warm periods, bees can continue collecting nectar and pollen from sources such as ivy. Most of this pollen is either stored for later use in the spring or it is eaten by the younger bees. The colony may well need also to collect water, although it will utilize any condensation formed within the hive.

When the winter cluster is fully formed, it appears as a ball of bees separated by the frames. The bees on the outside are the coolest at around 8 °C; in the centre of the brood nest, the temperature is maintained at 21 °C. We already know that individual bees will die slowly when their temperature falls to around 5 °C. The fuel used to maintain the temperature is the stored honey. These temperatures are only approximate. They vary with the outside temperature, the size of the cluster and whether or not brood is present.

Some colonies produce a little brood all through the winter while other colonies stop brood rearing entirely. The lengthening days of spring are one of the stimuli for the bees to prompt the queen to start to lay or increase her egg-laying rate. At first, the production of new bees does not equal the death rate of the older bees. This death rate is accentuated because the workers are older and the demands on their vitality are greater. Any of the bee diseases that affect adults add to this stress, so problems such as nosema, acarine and varroa (see Chapter 8) will accelerate the death rate. The colony can dwindle and may die. The population may just decline but, eventually, the number of new bees produced will be greater than the number of those dying.

Spring build-up

In spring (the time of which will vary from area to area), the population begins to expand rapidly so the graph of population growth in a colony can be shown as an almost vertical line at this point. Around this period, the queen lays the first eggs in drone cells, a while later, the colony makes 'queen cell cups'. These are the bases of potential queen cells and look very similar to the cups at the base of acorns. At some point in late April or early May, some will be prepared for the queen to lay in. All such cells prepared by the bees look polished to the human eye, just like the prepared brood cells described previously.

The queen lays eggs in queen cells in small batches so that the resulting new queens do not all emerge at once. It is important to know the timing of the development of queens, workers and drones and these are laid out in the table overleaf.

Queen cells point downwards with the opening at the bottom and are roughly conical in shape, often with a rough exterior. The egg hatches and the resulting larva grows until it spins a cocoon; the queen cell is then sealed by the bees with a wax capping. On the day, or the day after, the first queen cell is sealed, the colony is likely to swarm.

Days	Queen	Worker	Drone
1	egg laid	egg laid	egg laid
2			
3	3 days	as an	egg
4	hatch	hatch	hatch
5			
6		diet changed	diet changed
7			
8			
9	sealing	sealing	
10			
11			sealing
12	5th moult		
13			
14		5th moult	
15			
16	emerges		
17			
18			5th moult
19			
20			
21	mature		
22		emerges	
23	mates		
24			
25			emerges
26			
27			
28			
29			
30		flies	
31	begins to lay		
32			
33			
34			
35			
36		mature	
37			
38			mature
39			
40			
41			
42	too old to mate	foraging begins	

Life histories of the queen, worker and drone

The swarm

The swarm is a division of the colony which can be regarded as a type of reproduction. At the end of any particular colony's swarming period, it will have produced a prime swarm and maybe one to three casts, or even more. Without assistance from the beekeeper, only two or three of these are remotely likely to survive.

Prior to swarming, scout bees will investigate a number of suitable new homes. When the time to swarm arrives, bees appear to become agitated and fly out of their hive in great numbers. The queen has been starved or, more politely, dieted before this point. She becomes slimmer and lays fewer eggs. As a result, she is able to fly. It is rather like taking excess cargo off an overloaded aeroplane.

Up to half the colony population leaves in this first swarm which beekeepers know as a 'prime swarm'. Usually, the swarm bees cluster in a large mass quite near to the hive, within 10–20 metres. Swarms can weigh anything between 1.5 kg and 5 kg. The larger ones are often the result of two swarms emerging from different colonies in the same apiary at the same time. One of Herbert Mace's books, *Bee Matters and Masters,* has a photograph of a 6.5-kg swarm. It must have come from at least three colonies. If two colonies near to each other swarm at the same moment, their bees mingle in the air and are very likely to cluster together to produce a 4.5-kg 'monster' swarm. Large swarms from single colonies probably have a maximum weight of 2.5–3 kg. There have been various estimates or calculations as to the number of bees in a kilo. Results vary because when a swarm leaves a hive all the bees in it have crops full of nectar. The low end of the range is about 6,000 bees per kg and the high end (obviously with hungry bees) is 10,000 bees per kg. This means that Mace's 6.5-kg swarm could have had 42,000 bees in it or a very hungry 70,000 bees. About 40 per cent of the weight of a swarm is food being carried inside the individual bees.

Many people have actually calculated the population of a colony of bees, usually by weighing a selected sample and then weighing the complete hive with and without bees and calculating the population from the results. One researcher, Dr E. P. Jeffree, working in Aberdeen, calculated that his largest colony had a population of 49,000 bees (see Taking it further). I think claims of anything over 70,000 bees are as accurate as using the phrase 'a lot'.

Searching for a home

While the cluster is hanging, the scouts reach a consensus on where the swarm should live. They check out each other's discoveries and eventually a choice is made. The communication of direction, distance and desirability is by the bee dance (see Chapter 6). Once a decision has been made, the swarm takes to the air. The bees which 'know' where to go fly forward, through and around the swarm and this gives rise to descriptions of swarms circling and swirling. When in flight, the cloud of bees can be distributed, for a large swarm, over an area of around half a football pitch, but the whole clustered mass can fit easily into a 50-cm cube.

Fanning

On arrival at their chosen site, there may already be scout bees there, 'fanning'. These bees will be standing facing inwards towards the new nest site. They bend down the last segment on their abdomens to expose a membrane between the last two segments of the abdomen. The product of the Nasenov gland (see Plate 11) flows onto this and the bees blow the scent off into the air by flapping their wings vigorously. This scent is highly attractive to bees and draws the other members of the swarm, including the queen, to where it is strongest. Once close enough, at around one metre or so, they can see the mass of bees at the new entrance, head for it and run inside. If the entrance is small, it may take the swarm some time to enter. Bees like to choose a cavity with an entrance hole about the size of those that we like to put in bird boxes. Bee hives containing used comb are also very attractive to swarms. If you want to set up a 'bait hive' like this, it needs to be 2–3 metres off the ground with the used brood comb placed just inside the entrance.

The bees enter and start to cluster, hanging at the back and top of the chosen cavity. Contrary to popular belief among some beekeepers, most bees in the swarm are young workers. However much a swarm appears to increase in numbers, and they can appear to grow in size very rapidly, to start with the population is actually declining. The number of bees cannot begin to increase until the first eggs laid by the queen start to emerge as adults. As you now know, this takes three weeks. In order to live until then, the population of a swarm has to comprise mainly young bees. They start with a reserve of food, carried in the workers' honey crops. Comb is built as quickly as possible. As soon as some cells are completed, the bees regurgitate some of the nectar in their crops into the cells. As more cells are built, the queen starts to lay. It will, therefore, be three weeks before the numbers of bees in the swarm starts to increase. Up to that point, all swarms lose bees.

Swarms work very hard and collect as much nectar as they can. They are unlikely to swarm again that year but it can happen, especially if the swarm is a very early one.

About a week after the first or prime swarm has left the original colony, virgin queens start to hatch on the sixteenth day after the egg was laid. Then secondary swarms can begin to emerge, each led by a new virgin queen. If the colony was a large one and/or there was a large number of queen cells, there can be a number of such afterswarms or casts. There may be just one unmated queen in a cast or there could be several. The casts emerge a day or two apart and become smaller each time. I have collected some that could barely fill a pint glass.

The prime swarm will have to work hard to survive, but it does have the advantage of containing a laying queen. The casts will have to work harder still. Casts which emerge containing several virgins allow them to fight until only one remains. Then that one has to fly as described earlier, and mate with drones. A swarm can rear new bees in three weeks but a cast has to wait at least another ten days. About 50–70 per cent of young queens mate successfully. If this fails, the new colony is doomed. Experience has shown beekeepers that swarms either need to be fed or to benefit from sunny, warm weather in order to establish themselves quickly.

The original colony eventually allows one of the emerging virgin queens to damage any remaining queen cells.

The task of each of these colonies is then to collect enough food reserves and produce enough bees to enable it to survive the winter. It is during the winter or in early spring that most colony fatalities occur.

It is into this annual cycle that the beekeeper has to fit. In spite of what many people think, bees are not truly domesticated but both they and we benefit from good beekeeping.

03

equipment

In this chapter you will learn:
- the beekeeping associations
- about the equipment for you and your bees
- about the bee space
- about beehives and their parts
- about food and water
- about frames.

Beekeeping suppliers can supply everything you need to get started. Hives can be bought new or second-hand and bees can be acquired expensively or more cheaply. Always check with experienced beekeepers because quality can be independent of price and a good reputation is everything. Second-hand equipment will almost certainly be obtainable locally. Good beekeeping equipment can have a long life so, provided it is in good order, second-hand equipment can be very valuable. Not only should the materials be sound but you need to check that things have been constructed properly. I am using hives that had already been in use for 30 years before I acquired them. They are a bit battered and do not have the smart appearance of new ones but there are just as effective.

Free bees, i.e. swarms, or cheap unwanted colonies can also be acquired and, provided they are healthy, they can be requeened if you do not like their characteristics such as bad temper. To requeen a colony, the old queen is removed and replaced with a chosen one which can be purchased or reared yourself. Worker bees will kill a new strange queen by forming a ball of bees around her. Your new queen will therefore have to be protected when you introduce her to her new colony. Beekeepers have long understood that in order for the workers to 'get used' to a new queen, they need to be able to make contact with her but be prevented from being able to physically attack her. To enable this, the queen is placed in a cage made from a close mesh, allowing the workers outside to lick and feed her but not get close to her.

The whole process can go wrong and, unfortunately, bees have no respect for the amount of money you might have spent on your new queen. I strongly suggest that you get help from a local experienced beekeeper, maybe the same one as reared the queen. Done properly, the process has a success rate of over 90 per cent. It can have a failure rate of over 90 per cent if you do not understand the process. As a new beekeeper, get some help. Within a few weeks, the new queen will completely change the colony as the eggs that she lays hatch and the workers from the previous queen die off.

Beekeeping associations

Beekeepers, on the whole, like to help. It pays the new prospective beekeeper to join the local beekeeping association. Not only will the meetings give you a chance to meet other, more experienced beekeepers who can help you when you start, but many associations run courses for beginners as well. Through association membership, the beekeeper has potential solutions for any problems that may arise from his or her beekeeping activities. There should also be third-party insurance cover available to help with possible legal problems (see Taking it further).

The British Beekeepers' Association (BBKA) organizes an annual Convention each spring where there are helpful lectures and workshops and also the opportunity to buy equipment from many beekeeping appliance manufacturers from both home and abroad. Certainly for the beginner, this event is well worth attending because everything that you will need can be seen, ordered or bought on the spot. Confusingly, everyone you ask will give you advice at great length and much of it will appear contradictory. Follow one line of advice until your growing experience allows you to pick and choose with confidence.

The beekeeper's personal protection

The main component of a beekeeper's equipment is a good veil (see Plate 12). Today, the best veils come in the form of a combined jacket and veil or even a veil incorporated into a full set of overalls. Correctly worn, this type of garment minimizes the risk of getting stung. The downside is that they allow beekeepers to tolerate bad-tempered bees. Don't feel too immune. Bees stings can penetrate clothing, especially if the fit is too tight.

The new beekeeper will also need gloves. The more stingproof the gloves, the more clumsy they make handling frames. The best supple leather gloves with gauntlets covering the forearms are expensive. You must make your own choice. I favour a reasonable compromise and use household rubber gloves. I only wear them when bees are stinging freely. They can be thrown away when they split or get dirty. Bees won't just confine their stings to the beekeeper. We must keep bees that our neighbours can tolerate. Remember – they will not be wearing veils!

You will also need a smoker and a hive tool. The smoker needs to have a wide barrel and one made from stainless steel is very suitable (see Plate 13). Copper ones are good too and are just as durable for beekeeping as ones of stainless steel. Tinplate smokers quickly lose their plating and begin to rust. In spite of this, they can still be used for many years. However, modern stainless steel is cheap and will last for a long time so it is worth buying as good a smoker as you can afford. Smokers with wide barrels are easier to light. Before you buy, make sure that a good jet of air comes out of the nozzle when you pump the bellows.

Hive tools

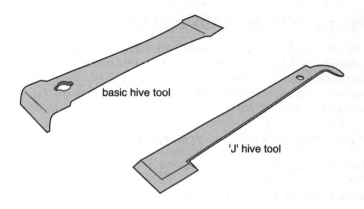

basic hive tool

'J' hive tool

figure 4 basic and 'J' type hive tools

A hive tool is an absolutely essential piece of equipment. There are two main shapes, both of which are effective in use. The 'J' hive tool has a 'tail' which is used to lift out the first frame. I like this one. The curved tail is slipped down between the frame and the side of the hive, i.e. in the bee space. The little step opposite the tail sits on the next frame. When you push the top of the hive tool away from you, the 'J' tip will push up the frame from below. Once this frame is out, the other end of the hive tool comes into play. Grip the 'J' end and you will see that there is a projecting free edge at the other end. Holding the hive tool horizontal, insert this edge into the bee space between the next two frames. Move the 'J' end towards your body and lever the frames apart gently. If you are right-handed, separate the left end first, then the right. This minimizes the movements of your hands across the frames and means that your two hands are then ready at the lugs to lift the frame out of the box.

The other type of hive tool is a more uniform width and has a sharpened blade at one end. The other is curved over so that it can be used as a scraper.

If possible, see both types in use before you buy the one you like. When you use it you will know if you have made the right choice! When you are inspecting your bees, get into the habit of holding the hive tool at all times. Hold it in the palm of your hand with the ring and little fingers. It is then instantly available when you need it.

Hives and the bee space

All modern hives are based around the discovery of the bee space by the American, the Revd L. L. Langstroth. When bees build free comb in a box or skep (straw beehive), the lower one- or two-thirds are free hanging and not attached to the sides. Langstroth made frames to hang freely from projecting top bars which could be suspended in a box. He contrived to have the natural space bees left around the lower area of the comb around each side of the frames, right up to the top bar. He found that bees accepted this very readily. He also found that he had built 'moveable' frames. This discovery has allowed modern beekeeping to develop. Other people had previously tried to use frames but without the surrounding bee space. A German, Dzierzon, had also made a moveable frame around the same time, but his frames stood rather than hung freely. Langstroth's discovery was much more practical; the beekeeping world is indebted to him.

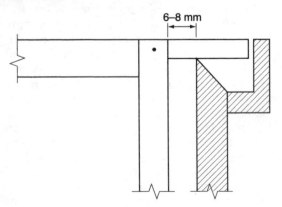

figure 5 the bee space in a moveable frame hive

The bee space measures around 6–8 mm. Worker bees, queens and drones can all pass through the bee space so it is incorrect to say that the bee space is the space a bee can just pass through. Workers need much less space to pass than queens or drones and this fact is used in the design of the queen excluder. In British hives, this is usually made from a thin sheet of metal out of which slots are cut that are big enough only for workers to pass through, or a wire grid with similar sized spaces between. The overall size of the queen excluder corresponds to that of the hive cross-section. When the queen excluder is placed above the box containing the brood, it serves to keep the queen and drones below while allowing workers access to the other parts of the hive above the excluder.

The bee space is not an absolute constant. Bees naturally build different bee spaces in different parts of the hive. Natural combs are built to within two bee spaces (about 13 mm) of the bottom of the cavity. The combs are built with a relatively equal space between the midribs of adjacent combs. In natural colonies, this space is roughly constant but is not an invariable measurement. However, modern frames and modern hives are designed around a fixed measurement. This is 35 mm in American-style hives or 38 mm in some British hives.

Frames can be spaced at the appropriate distance apart by using either a metal or plastic spacer which is slipped onto the lug.

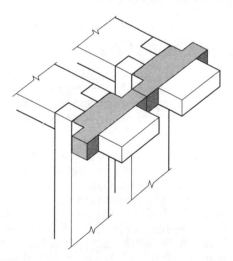

figure 6 removeable spacers or metal/plastic ends

This type of spacing only really works on the British Standard frame which has a lug 38 mm long. A much firmer spacing is achieved by widening the top part of the frame's side bars. This type of spacing is found in the Hoffman frame and is probably the commonest frame spacing method worldwide. Bees stick frames together where they touch and they stick frames to the hive body at any point of contact. They stick hive boxes and the excluder together. To reduce this, all these points of contact have to be minimized. The Hoffman frame has a 'V'-shaped vertical edge at the top of its side bars which reduces the frame-to-frame contact area.

Where honey is stored in the colony, bees build out the cells until the space between the face of one comb and that opposite it is one bee space wide. Where brood is being reared, the space is twice as large, i.e. two bee spaces wide. In my experience, 38 mm spacing in the brood nest is just as effective as 35 mm. I actually prefer the wider spacing because this was the one around which British hives were originally designed. This means that the Modified National, the commonest British Standard hive, was designed around the 38 mm spacing. Ten frames fit into the brood box of a WBC hive (described later in this chapter) and 11 into the Modified National brood box. When you use the 35 mm spacing that is commonly available and used in the USA, it leaves a gap at one side of the hive. The best solution in this situation is to fill this gap at the outside edge of the box with a dummy frame.

Top or bottom bee space

Horizontal bee spaces in artificial nests (hives) need to be accurate to be well accepted by bees. British Standard frames hang in their boxes leaving a bee space between the bottom of the frames and the bottom of the box that they are in. They are known as bottom bee-space hives.

The long lugs of British Standard frames are covered by fillets. The bottom one covers the frame lugs and creates a bee space right to the end of the frames. The top fillet creates the cavity in which the lugs rest. The frames hang so that the top bars are level with the top of the box and they rest on inverted, v-shaped metal runners. The fillets in the Modified National hive create wonderful hand holds and the ends of the box at right angles to the fillets create vertical hand holds so that, if hives are carried in vehicles, they can easily be grasped there which makes pulling the hive to the handler very easy. Where the frames are arranged

Top bee space

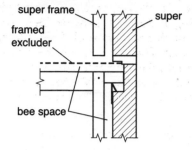

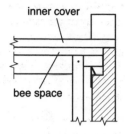

Bottom bee space

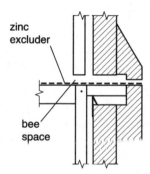

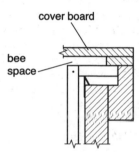

figure 7 top bee-space and bottom bee-space hives

to hang with the inter-box space above the top bars, the hive is known as a top bee-space hive. The prime example is the Langstroth hive.

British Standard frames are those found in the National, Modified National and WBC hives. As far as I know, National hives are no longer made. In these, the end walls where the frame lugs rested were double walled. Each side looked the same outside, with simple hand-holds like those on a Langstroth hive. The National was modified (hence the name) to hold the same frames but to make it single-walled using thicker timber. Today the modified version is the one referred to by beekeepers as the 'National' hive.

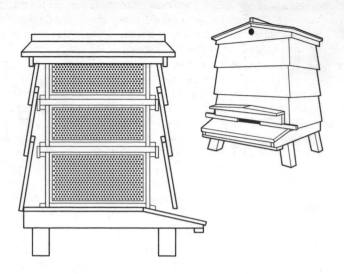

figure 8 the WBC hive

However, all hives follow the model created by Langstroth and this is the most widely used design worldwide. This is not the case in the United Kingdom. The Langstroth has frames with short lugs which rest in rebates cut into the thickness of the timber walls. It is also a top bee-space hive. Advocates of the top bee-space are convinced that it is superior to other bottom bee-space hive designs. However, once the boxes are put together on top of each other, I cannot see that there is any difference. In both the Langstroth and Modified National, the bee space runs the full length of the frame. It is only in the WBC that when other boxes are superimposed so that the edge of the upper box rests on the spacers of the frames in the lower box.

The WBC hive is named after its inventor, William Broughton Carr. In truth, there were many similar hives in use at the time. The hive we know today as the WBC was not the one originally designed by Carr but a modification made by the, now defunct, company, Lee's. It is a double-walled hive. Separate inner boxes hold the bees and these are covered by outer 'lifts'. It is an expensive hive to buy new but it can be very cheap second-hand.

The lifts are actually just large enough to cover Modified National boxes. People like the look of these hives, so if you are using Modified Nationals, you could place them inside a set of WBC lifts to achieve an attractive effect. This is the hive most generally recognized as such by members of the public. The sloping lifts and gable roof give it a pleasing shape. The stand incorporates a small strip or alighting board in front of the entrance where the bees can land before walking into the hive. It is the smallest hive, holding only ten British Standard frames and is often decried as being too expensive and too fussy. In my opinion, they use too much wood and are slow to work with. However, they are very weatherproof which may be a consideration in your situation. I also have to say that the highest average crop I have heard of for the United Kingdom was obtained from bees kept in WBC hives. Two tonnes of honey were collected from 17 colonies. This works out at over 115 kg per colony.

The disadvantage of using American hives in the United Kingdom becomes apparent when you try to sell bees and equipment. Comparatively few beekeepers use them. However, they are just as good as any other kind of hive.

All hives, in spite of what their advocates say, have disadvantages. All designs are compromises between the perfect piece of equipment we all want and what is reality. My advice is to have only one type of hive. Do not mix them and do not change patterns until you have really thought about it.

Brood boxes and supers

Most hive types have boxes in two sizes. In reality, the only difference is in the depth of the side walls. The brood box in Modified National hives has to hold the British Standard frame. This is 210 mm deep plus the bee space below that. This means that the total depth is 225 mm deep. The shallow box is meant to hold shallow frames from which honey can be extracted, and which are 140 mm deep. I find shallow frames much easier to handle and uncap. All boxes, brood and super, are 460 mm square.

However, many beekeepers use deep boxes both for brood and honey supers. I have known a few beekeepers who kept their colonies solely on shallow boxes. There is no rule to say what you must do, but before you invent your own hive, do find out how everything ticks. There is nothing easier than inventing a hive that doesn't work. However, bees are so adaptable that even weird hives can actually work.

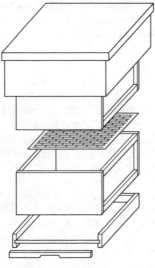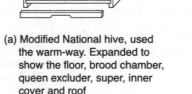

(a) Modified National hive, used the warm-way. Expanded to show the floor, brood chamber, queen excluder, super, inner cover and roof

(b) Modified National hive, used the cold-way

figure 9 the Modified National hive showing frames positioned 'warm' way and 'cold' way

What you will need

If you stick to basic methods, you will need a certain amount of equipment for each colony. If you start with full colonies early in the spring, you will need it all before very long. From the ground up, you should have:

- one floor and entrance block or one floor using the shallow side
- one (or two) brood boxes with frames and foundation
- one queen excluder
- three or four supers with frames and foundation

- one inner cover
- one feeder
- one roof

You will also need a sturdy hive stand on which to put your hive. Remember that this will have to support a hive with a number of heavy supers if the season is kind.

Apiary sites

You must think about where your hives will go. Very little space is required as regards area but there are many other considerations. Modern gardens are so small that your bees, wherever you put them, will be not only near your house but also near those belonging to your neighbours. If you are in this situation, it is probably better to find a site in a corner of a field if you can. Such out-apiaries must be fenced to keep livestock out. If you can find an area at the edge of woodland, in general, this can be very suitable.

If your garden is large enough, site your hives where they are least likely to be seen. They can be kept inside a building or a shed. In Germany, hives were traditionally kept in bee houses and they still are in countries such as Slovenia.

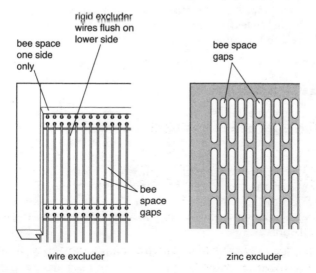

figure 10 wire and zinc queen excluder patterns

I have seen two hives in the attic of a three-storey house. The flying bees used the skylight and the colonies were doing well. The worst factors in this particular situation were the poor light levels for when the beekeeper was inspecting his colonies and climbing the stairs to look at the bees!

Gardens which abut onto fields make very suitable apiary sites. The secret of a good site is not allowing problems associated with the flying bees to occur. Hives should be placed where the bees have to fly up and over obstacles and over the heads of passers-by. Bees perceive fences made from small mesh wire netting as barriers and fly over them. Sites in the centre of a group of shrubs also work very well. If you get complaints about bees in your garden, you must make some effort to move your colonies. People have a right to 'enjoy' their property.

Apiary sites should be sheltered from the wind. The best shelter here is hedges. A thick hedge will slow wind for a distance along the ground equal to 40 times its own height. Apiary sites can be too hot so deciduous tree cover allows both summer shade to keep the colonies cool and exposure to the winter sun when they will appreciate the warmth. Face your hives away from the prevailing wind so that the returning flying bees have to approach upwind which makes manoeuvring easier. Bees quite like to have their entrances facing south but in a sheltered apiary site I don't think this matters too much. If your apiary is away from home, it is quite useful to be able to drive right up to the hives. I quickly learnt that a light super at the start of a 50-metre walk weighed a tonne by the end of it!

Water supply

Wherever you establish your site, you must make sure water is available. The only complaints I have had about my bees where I now live have been about them drinking from garden ponds. Bees love such sites. If the bottom of the pond slopes, it means there is a shallow edge which will warm up much quicker than the bulk of the water and bees will collect there. They like to collect on the edge of the pond, on lily pads or duckweed. You may well have a pond of your own but bees like to drink from several sources.

If water is not readily available near your apiary, you can set up a water fountain (see Plate 14). The simplest design you can make is to stand a large container with holes in the bottom in a bucket which you must keep topped up with water. Fill the container with potting compost, peat or something similar which will draw up the water and present a damp surface to your bees. Don't set out just one – set out half a dozen and stand them in the sun. You can also use a bowl of water containing pottery crocks, stones, etc. to give the bees something to perch on.

To 'train' your bees to your new water fountain, feed your colonies with some syrup containing a little peppermint essence and fill the water reservoir or bowl with dilute syrup which also contains the essence. The bees will find this and, as they empty it, you can simply top up the reservoir with ordinary water. Another possible water source is a slowly dripping tap, particularly if it drips onto inclined planks.

It has to be done this way. Giving bees water in the hive simply doesn't work. Bees seem to like water provision to be at least six metres away from the hive. Watery syrup will work if you place it directly on the hive but it has to be at a concentration of at least ten per cent sugar to attract even starving bees. If your tea was this sweet, most of you would hate it.

The catalogues of appliance manufacturers are full of all sorts of equipment. Some is very useful and some useful only in the mind! Buy the basics to start with. The ephemera are exotic and can be purchased as the mood takes you.

Other hive types

There are a number of hives that use frames with short lugs. I have already mentioned the Langstroth; others are the Dadant, the Jumbo Langstroth, the Smith and the Commercial. The Dadant and the Jumbo Langstroth are large hives. The Smith hive will take British Standard frames except that they need to have a short lug. The Commercial is also known as the '16 × 10'. Its frames measure 16 inches by 10 inches (approximately 400×250 cm). The actual cross-section of the box is compatible with that of the British National hive so boxes, floors, inner covers and roofs of both types can be used together but the frames cannot be interchanged.

On the Continent, even more different hive types are in use but all have the common use of the hanging frame and the bee space, thanks to Langstroth.

Frames and foundation

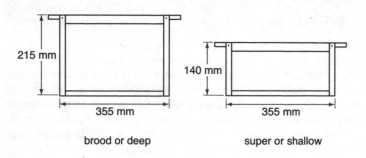

215 mm — 355 mm — brood or deep

140 mm — 355 mm — super or shallow

figure 11 British Standard frame dimensions

Frames can be nailed together days, weeks or months before use. However, you should not insert foundation into them too soon or it will tend to go brittle and buckle out of shape. Personally, I don't think substitutes like plastic foundation and semicomb are worth the cost. These days, foundation is generally available in packs of ten sheets. That sold in England for use in brood frames is thick enough for eight sheets to weigh one pound or 14 shallow sheets for the same weight. In these days of metrication this is just one more remnant from earlier days.

Even at this thickness, the wax sheet will stretch when bees cluster on it to build comb. In order to be attractive to bees, the shape of the cell is embossed on the sheet. The bases of the cells in early foundation were flat bases but the bees did not like this. Cells in modern foundation have dimpled bases made up from three parallelograms, like natural comb.

To stop foundation stretching, it is strengthened with wire. Various methods of wiring have been used but most foundation today has a zig-zag pattern with loops of wire projecting beyond the edges of the wax sheet. The wire is stretched taut over the wax sheet and an electric current of 12 volts or so is passed through. The wire heats up and melts into the wax. However, wire and wax expand and contract at different rates which means that if the foundation is put into the frames too early, over time, the wax will develop a series of waves. The resultant comb will therefore not be flat. This is an essential requirement for easy colony management.

In storage, wax develops a bloom which resembles the bloom on dark grapes. At the same time, the foundation can also develop the same wavy shape as before – it does not have to be in the frame to do this. To avoid this, I suggest the best way to store foundation is on a flat surface. Make a neat pile and place another flat surface on top. Two sheets of plywood will do the job. Put a light weight (a thick book you never read, for example) on top. The flat surface stops the waviness. The whole ensemble should be kept in a warm place (I use the airing cupboard). Just warm is warm enough – hot is too much and you will be clearing up a very messy situation! Warmth seems to stop or delay the formation of the bloom and leaving the foundation in its plastic bag will mean it will keep very well. Figure 12 shows the usual wiring pattern.

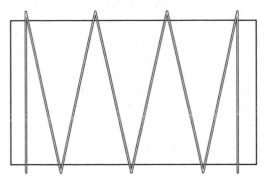

figure 12 wired brood foundation

The longer wire loops go at the top when you put the foundation in the frame. They are bent through 90 degrees and fitted into the angle in the top bar made when the wedge is removed. The wedge is then put back and nailed in place.

You will find it useful to assemble frames and fit foundation in the same sequence each time. Try to assemble and nail the frame together without removing the wedge in the top bar. This simply means that the wedge is handy when you need it. A top bar always has lugs that are 22 mm wide and 9.5 mm thick. The joint where the side bar fits on is, for the British Standard frame, always 9.5 × 9.5 mm. It is the weakest part of the frame and if breakages occur in use, they are almost invariably at this point.

Frame side bars generally have a groove on the inside face. The foundation edge slides in here so that the whole sheet is held straight. If the groove is inside when you assemble your frames, all will be well. You have to try hard to get it wrong! For Hoffman frames, if the grooves are facing inwards, the 'V' shaped edge will always point forward on the right-hand side of the frame and backward on the left-hand side.

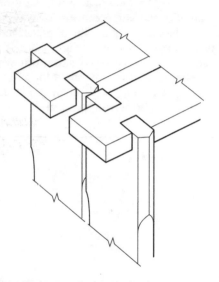

figure 13 Hoffman self-spacing frames

The bottom of the frame is in two pieces (the bottom bars), one to go each side of the sheet of foundation. Nail all the frame joints except for one of the bottom bars. Just before inserting the foundation, remove the wedge and cut away all traces of the supporting shim. If the frame is nailed up completely, it is quite possible to slide the foundation between the bottom bars, then bend the wire loops through 90 degrees and make sure the foundation fits snugly into the recess in the top bar. Don't force it. You may need to remove a sliver of wax from the side if the sheet is too wide to slide in easily. Some people think it is easier to insert the foundation with one bar removed and then to nail that in place afterwards, but you have to take care not to buckle the foundation when inserting this second bottom bar. If the frame has been nailed together correctly with all the angles at 90 degrees, the foundation should fit well. Fold over the loops and press the wedge tightly into place. I nail it in place with four nails. In my opinion, there is no need to make sure any of the nails go through the wire loops as the foundation is held in place by pressure from the wedge.

In most British catalogues, deep brood frames with straight side bars are designated DN1 and the shallow super frames SN1. Hoffman self-spacing brood frames, which are wider at the top of the side bars, are designated DN4, with the equivalent super frames being SN4. Top bars can be either narrow or wide. If you use the wider ones, this makes the space between the top bars about one bee space. This reduces any tendencies bees might have to build comb between the frames. I expect similar methods to be used in other countries.

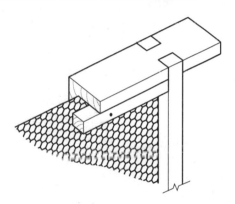

figure 14 fitting foundation into a frame

Frame nails are known as gimp pins and are like panel pins with a small head. This allows them to be removed easily. Once a comb has had its day, the wax can be removed and the frame used again. When you lever out the wedge (using the hive tool, of course), the head means that the frame nails are pulled out with it. Scrape everything clean, particularly the side grooves and between the bottom bars. If everything is clean, you can slide a fresh sheet of foundation into the frame and replace the wedge. This means that a frame can be used over again. Drive the nails in with a light hammer or push them in with the appropriate push-pin tool.

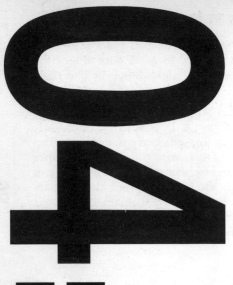

04

beginning your beekeeping

In this chapter you will learn:
- how to acquire bees
- what they need
- about your friend the smoker
  – and how to light it
- how to help your bees to develop
- how to examine a colony
- what you are looking at
- about stinging and bad temper
- about seeing the queen – or not.

Obtaining your first bees

Bees can be acquired in various ways. If you join your local association, this opens up a number of possibilities. Most associations produce a magazine or newsletter in which you may find advertisements offering bees for sale, or you can place an advertisement asking for bees. The time to acquire them is after the early spring. The most likely time for bee colonies to die is February–March, so buying from mid-April onwards should give you a good chance of buying a survivor.

What to look for

1 Visit the colonies during the day in sunny weather. Bees should be flying well from the entrance.
2 Check that the bees are bringing good-sized pollen loads home, i.e. they should be very obvious to the naked eye.
3 Observe what the bees are doing. Are they taking an interest in you and the beekeeper? You should be able to stand behind a hive for a long time without being bothered by the resident colony.
4 Look at the equipment if this is included in the deal. The woodwork can look scruffy and still be sound. Make sure the measurements are correct (at least the internal ones which are the important ones in this case) and make sure they are of the type you have chosen. Check whether the roof leaks and make sure all the parts fit together and are bee-tight.
5 Look at the bees when the hive is opened. Make the beekeeper do it if you don't feel confident. Are the bees docile or are they trying to sting? This is important when you are considering where you intend to keep your hives. Docile bees can be kept virtually anywhere. Bees that are rather more bad tempered may be fine in an out-apiary and you can requeen them if you wish.
6 Look at the combs. If they are really black, they are old. Look at the broodnest. The brood should be normal. Take an expert with you if you are not sure. Brood combs should be covered thickly with bees. If you can see the brood patches easily because the bees are not covering them, it may mean the death rate among the adult bees is very high. In which case, don't buy.
7 Ask about the history of the stock. It should be similar in size to the others in the apiary. If it is smaller, it could be diseased. If it was a nucleus in the autumn, then this could account for the smaller size and it would be a good buy.

8 Look at the food reserves. If they are low – equivalent to around two brood frame – you may have to feed.

9 The really important points are:

i the normal appearance of the brood patches and that they do not have lots of missed or empty cells among the sealed brood;

ii the docility of the bees themselves;

iii the interchangeability of the equipment you are buying with any you may already have;

iv the age of the queen – she should be young.

Suppliers of bees and equipment

All beekeeping equipment suppliers also sell bees. Nationally distributed magazines carry advertisements for bees and queen bees during the active season. You will pay more for your bees this way but you should be dealing with a supplier who is more aware of his reputation and someone more able and willing to rectify mistakes. Take your time. It is much easier to buy poor quality bees than you think.

Bees should not be disturbed during the dormant season, another reason for acquiring your first colony from April onwards. Remember, colonies that are strong then will soon require supers and regular examinations and you will need the full set of equipment (listed in Chapter 3). To be effective, the person handling a large colony really needs experience as swarming may well start around this time. I therefore recommend that you start with a nucleus which is a small colony, on four or five frames, with a new, young queen. Bought or acquired between May and mid-July, it should at least build up to be strong enough to live through the next winter. By then, such a nucleus should be the size of a full colony – and you, the new beekeeper, will have gained experience. With luck, you may even get a little honey.

Letting the bees out

If you have bought bees from a supplier, they usually arrive in a non-returnable travelling box. If you have acquired them locally, they may be in a nucleus box which you will have to return. They should already be on the type of frame that fits into your chosen hive. You may have bought them with their hive, thus avoiding the need to transfer the colony into your own equipment. Whether or not you have to return the container in which they arrived, your first actions will be very similar.

Place the hive or box on the site where you want your colony to stand. You need to keep the hive off the ground so that it isn't affected by the damp earth. It is convenient to have the top of the brood box at a level that means your hands will fall naturally to handle the frames. Many beekeepers end up suffering from 'beekeeper's back' because of spending time bent over a beehive or because they do not lift heavy weights, like full supers, properly, using their knees rather than their back muscles.

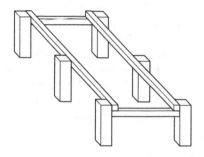

figure 15 a hive stand

Hive stands should be level so that comb is built within the frames which will then be vertical. The simplest is made from two 450 × 230 mm concrete building blocks and two 50 × 75 mm wooden rails cut to about 1.5 metres. Nail two pieces of wood across the ends of the rails so that they are about 350 mm apart on the inside measurement. If the rails are 50 mm thick, the outside measurement will be 455 mm. Support this oblong frame with one of the blocks at each end and use packing if necessary to make sure it is level. This length is sufficient to support two hives which will make manipulations such as the artificial swarm easier to undertake. Whatever you use as a stand, it should be strong and stable. The hive should be a minimum of some 300 mm off the ground.

Transferring the colony

Face the entrance of the nucleus in the chosen direction. Leave it standing there for a little while and put on your veil, gloves and any other protective clothing.

Some travelling boxes are made of stiff cardboard or thin plywood designed for single use and you may only have to push the entrance open or prise off a closure block. The entrance closure may be tacked onto the box or the entrance could be blocked with plastic foam, similar to that found in the cushion you are probably sitting on at the moment. Make sure you have any tools that you think you may need, or you could use your hive tool – that is what it is for! As well as facing the nucleus in the required direction, try to make sure that the entrance of the box is at the same height as that of the hive which will replace it.

Boxes in which bees travel should have lots of bee-proof ventilation. Before you open the entrance, cover this ventilation with a cloth to prevent bees returning to the ventilated screen rather than going in at the entrance. Something like a sack will be sufficient, but make sure you don't cover the entrance itself. That done, open the box. Bees may well pour out if they have got hot, or they may emerge more sedately.

Remember that bees become part of the flying population when they are middle-aged and form only part of the total colony, so your nucleus won't all fly away. Flying bees remember their hive entrance as a point in space. Your bees, whether they pour out instantly or emerge slowly at first will fly in circles, first one way and then another. They may well be a bit annoyed from being shut in, which is why you put on your veil. Soon, some bees will find the entrance and settle. Some of them will stop at the entrance, lift their abdomens in the air and expose a light-coloured membrane between the last two segments of their abdomens. They beat their wings but don't fly. This blows a scent from the membrane into the air which attracts bees to its source. In this way, bees learn the location of their hive.

I suggest that you allow the bees to fly like this for at least an hour. Better still is to leave them until the next day. If rain threatens, cover their box with something waterproof such as a sheet of plastic, but make sure that it cannot blow away and that it doesn't block the entrance. If need be, provided all is dry and secure, the nucleus can stay there for several days – at least until the next fine day.

Lighting the smoker

Before you go through your colony, you must light your smoker.
Like anything else you may never have seen done, or done
before yourself, practise first. Suitable fuels must be dry. Very
soft, rotten wood is a good fuel but it can burn away swiftly.
Small pieces of this light easily and smoulder gently. Put a piece
that is smoking freely, about the size of a walnut or small apple,
into the smoker barrel. You can then top this up with planer
shavings, but don't compress them. Keep puffing away with the
bellows until you get a good stream of smoke coming out of the
nozzle. Then you can leave it with just an occasional puff to
keep it going. If you have lit the smoker properly, it should still
be capable of producing smoke when you need it half an hour
later.

You can also light the smoker using newspaper. Screw half a
tabloid-sized sheet up loosely, light it and put it in the smoker.
Keep puffing and gradually add shavings a pinch at a time. The
object is to build up a bed of hot embers at the bottom of the
smoker.

Other suitable fuels include egg boxes, dried leylandii leaves,
jute string, dried lawn cuttings or sacking. Many modern
cardboards are treated with a flame retardant and are not
suitable. If you use loose fuels, such as shavings, then put a twist
of green grass on top before you close the lid. This stops sparks
flying from the nozzle. I have seen a beekeeper panic and puff
so hard that he produced flames from the nozzle. Impressive,
but not what was required.

Having flown for a while, an hour at least, your bees can be
transferred into your hive. Blow a little smoke into the entrance
of your nucleus box and wait. I use this time to put on my veil.
Next, move the box to one side and place the hive floor in that
position. Remember – the hive must be level. Place the brood
box on the floor.

Warm way or cold way?

The Modified National brood box can be placed either so that
the frames are at right angles to the entrance (a) or parallel to it
(b). Frames at right angles to the entrance are said to be the
'cold' way while those placed parallel to the entrance are said to
be the 'warm' way (see Figure 9). To work the bees in position (a),

you need to stand at the side of the hive. For position (b), you need to be able to stand at the back. In either of these positions, your hands fall naturally to the frame lugs. Any other position makes handling frames awkward.

Check that your veil is on properly and the zip completely closed. It is also a good idea at first to put on your gloves. Those with long gauntlets fit over the sleeves of your overalls or bee veil. If you have sleeves that are elasticated at the wrist and you are using rubber gloves, I think it is better to tuck the gloves inside the sleeves. Often the sleeves of veils have an elastic loop which goes over the thumb and helps to keep the sleeve down. I find it less of a nuisance to cut it off. However, do try it round your thumb before you do anything as drastic. In beekeeping, nothing beats experimentation and experience.

When you are really ready, lift the cover of your nucleus box. You then need to lift out the frames in turn, using your hive tool to gently prise them apart if they are stuck. If you want your frames the warm way, put the first comb at the front and the rest behind, one after the other. Keep the same combs and comb faces adjacent to one another. Push the frames up to the hive wall just behind the entrance and then fill up the rest of the space with frames of foundation.

If you want the frames to run the cold way, the combs from your nucleus should go in the middle of the new brood box with frames of foundation on either side.

Closing up

Put the inner cover on, being careful not to crush any bees. Place it diagonally at first, with the rim downwards. This means that the cover is only touching the hive in four places. Then you can twist it gently to line it up with the box below. As you turn it, any bees that are in the way will move and you are less likely to crush any. Use a little smoke just to drive the bees down into the box.

Feeding

You need to feed your colony to help the bees build comb on the foundation as the little colony becomes stronger. The feeder can be as simple as a honey jar, or similar, with a few fine nail holes punched in the lid. Half fill the container with dry sugar and then fill up to near the top with hot water. Stir to dissolve the sugar. When you invert the container (with the lid on!), a few drops of liquid will escape until air pressure holds the rest in. You can then place the inverted container over the feed hole in the inner cover, exposing the holes to the bees. They will come up and suck the liquid from the holes. The first feed could be as large as five litres. You may have to rest the jar on slivers of wood 6–10 mm thick to give the bees space to reach all the holes. Refill the feeder as the bees empty it. An inspection a week later, if the weather is fine, should show the bees building comb or 'drawing out' foundation, as beekeepers put it. You will need an 'eke'. This is four pieces of wood nailed together into a box the same size as your hive. The pieces need to be the same depth and deep enough to hide the feeder.

Developing the nucleus

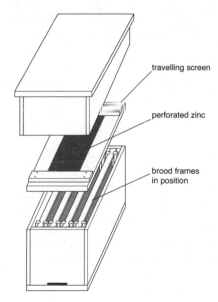

travelling screen

perforated zinc

brood frames
in position

figure 16 a nucleus box

The more bees there are in the nucleus, the greater the area of the brood nest that can be kept warm and the more eggs the queen can lay initially. The more she lays, the faster the new colony can expand. Honey can be collected in Britain between April and September, depending on your geographical location, but, in most places, unless your little colony fills the brood box before July you stand little chance of getting honey for yourself – but you will have learned something about handling bees and will have gained much expertise. I suggest that you, the new beekeeper, look into your colony regularly, every seven to nine days.

The later in the summer that you acquire a five-comb nucleus, the less likely it is to reach full colony size, but it can still easily be strong enough to live through the winter. Five, six or seven combs well-covered with healthy bees should generally do so.

The established colony

The examination of an established colony is slightly different from the transfer, but should follow the same basic steps. When you look through a colony, it should be for some purpose. The minimum reason should be to 'see what they are doing'. As a new beekeeper, this may be enough. However, the operation should never be frivolous, otherwise you could compare it with digging up a plant to see how the roots are getting on. So, first of all decide what your reason is. Between May and July, for example, it could be to see whether the bees are preparing to swarm.

The effects of smoke

First, having made sure that you have acquired the skill of lighting the smoker, you should blow smoke into the hive entrance. In my opinion, smoke is more effective than other methods of colony control. Smoke does two things. Firstly, it drives bees away from the area where it is applied. Secondly, it causes the bees to eat from open honey storage cells. The popular explanation is that the bees eat food in order to help the colony survive if they have to leave their hive. If there are no such open cells, it may take them some time to chew through the cappings, but most colonies require one or two minutes for the smoke to have the desired effect. You will hear beekeepers say that bees react as they do to smoke because of the fear of fire. Really wild bees, such as the giant bee (*Apis dorsata*) in Asia instantly fly off the combs when touched by smoke. UK bees do not. I suspect that, one way or another, we have unknowingly favoured bees that stay in hive and eat honey (engorge) when they are smoked. What a disaster it would be if they all flew away to escape the supposed impending disaster!

Bees that have eaten honey seem to be more docile and do not react as badly to the hive being opened. Some bees, if handled gently, require little or no smoke, but it is wise to use smoke when you first start inspecting colonies. Good weather – sunny and warm – is also likely to make the bees more docile, while bad weather – cold, windy and wet – will make them worse. As a rule of thumb, choose a day when you feel comfortable outside for the first occasion you open your bees. The bees will almost certainly be more comfortable too!

Opening a colony

Take off the hive roof. Turn it upside down and place it on the ground within easy reach. The edges of the roof can be used to support other hive parts, such as the inner cover and supers, placed on it at a cross angle. The weight-bearing points are small enough to make the prospect of crushing workers or the queen highly unlikely.

Remove the inner cover. The bees will probably have stuck it down with propolis, a resin they collect from trees for this purpose. Bees use propolis to varnish over surfaces as well as for sticking adjacent surfaces together. The bond can be very strong which is why you need a hive tool to separate hive parts. Hive tools combine the properties of a lever and scraper.

Using the smoker

Lift up the inner cover slightly and puff in a little smoke. Then remove the inner cover and examine the underside. This is to make sure that nothing important is on it. If there are no supers on the hive, the queen could conceivably be under the inner cover. She could also be on the underside of the queen excluder if there is one on the hive.

Marking the queen

Handling queens can be daunting for the new beekeeper. Buying bees from a known source may mean that the queen has already been marked for you. In other words, a spot of quick-drying paint has been put on her thorax. This makes her very much easier to spot. If you want to try this, practise on some drones. They are easy to see and don't sting. However, you *must* dispose of them afterwards. How do you expect to find a yellow-marked queen in a colony full of yellow-marked drones?

I suggest that you use what is known as a Baldock Queen Marking Cage. It is also referred to as a 'crown of thorns'. It consists of a circular ring with threads criss-crossing the middle and a row of spikes set in one face of the ring. Place this carefully over your queen and press the spikes down into the comb. The threads will hold her against the comb and she will not be able to escape between the spikes as they are set too close together. Ease the cage up a little to allow the queen to move so that you can frame her thorax within a square of the threads. Then press down again gently, just enough to hold her still but not divide her into cubes! Then apply the spot of paint to her thorax. You can buy queen-marking paint from the appliance suppliers or I use Humbrol® model paint. This comes in small tins and a variety of colours. The fluorescent yellow and orange ones stand out very well in the hive. Apply this with a small brush or the end of a matchstick. Give the paint a few moments to dry before you release the queen back into the colony. It is worth watching her until you know that she has been welcomed back by the bees. Sometimes the workers take exception to the smell of the paint and they can attack the queen. If this happens, break up the ball of bees surrounding her and hopefully all will be well.

There is an internationally agreed sequence of colours denoting particular years. Using this, you can easily see in which year your queen was raised and hence how old she is. Five colours are repeated twice over a ten-year cycle based on the final number of the year.

1	White	6
2	Yellow	7
3	Red	8
4	Green	9
5	Blue	0

There are many mnemonics to help you remember this sequence. I use the astonished 'What? You Rear Green Bees?'

If you are looking at a full-sized colony in April, the bees may fill the box, i.e. there will be bees on the frame tops right up to all internal edges. The queen is least likely to be on the frames at either end of the box. Choose the end frame nearest to you and use the hive tool to break the propolis seal by levering gently. You need to master the skill of keeping hold of the hive tool at all times. Use your ring and little fingers to keep it in place. If you are right-handed, hold it in your right hand, and vice versa. Bees may appear to be milling about everywhere. Brief puffs of smoke at the frame ends may be enough to move them away from where you wish to touch. Smoke not only causes bees to engorge before the hive is opened, but it can be used to move them away from certain areas.

What you need to notice

Lift out the first frame. The aim is to do this smoothly. Do not jerk or bang anything. The frame may be light and empty or full of food and covered with bees. Try to identify as many things as possible. Does the frame contain brood? Is it young or old – i.e. eggs and larvae or sealed cells? Are there pollen stores? How much honey is there on the comb? Make sure you hold the frame over the hive in case anything precious, such as the queen, falls off. Better in the hive than in the grass.

Put this frame down. It can be placed in another brood box or cardboard box. I stand mine on the hive stand. The 355-mm gap I advised when you were making the hive stand allows you to place the frames on the stand at an angle. One lug rests on the far rail and the bottom bar rests on the near rail. The lug you used to put it down sticks up in the air. The frame is now both safe and handy. You can also lean it up against the hive or the hive stand, resting on the lug and one bottom corner. Some beekeepers use a special box to temporarily house this frame.

You have now created a space in the box. You may need a puff or two of smoke to clear bees away from the lugs of the next frame. Then use the hive tool to move this slightly into the space. You can then lift it up smoothly to eye level. When you are examining a colony, you can stand up, kneel down or sit on something like a fisherman's stool. You will find which suits you best by experience. Examine the next frame and, when you replace it in the hive, move it back firmly into the space made by the first frame. This means you keep the working space into which you can manoeuvre the next one before you lift it out. The space moves with you as you work your way across the brood box.

Some beekeepers use cloths to cover the frame tops, just exposing the gap and the next frame to be examined. The cloths are attached to sturdy lengths of wood which are wider than the hive and thus able to rest on the hive walls. Place one cloth at the end of the hive, parallel to the frames. Roll it back to reveal the first frame that you plan to lift out and the top bar of the next one. When you want to move on to the next frame, lift it out and inspect it. When you have finished, replace it in the gap produced by the removal of the first frame. Then roll the first cloth back to expose the third frame. Place the second cloth, rolled up, over the frame you have just put back. Continue across the brood box until you have finished your inspection. I find cover cloths a nuisance, but try them if you want to. They do keep the bees down in the rest of the brood box.

Once you feel you have seen all you need, push the frames back into position en bloc or in groups of two or three. Do this with your thumbs or your hive tool. Give the bees a chance to move out of the way, or gently persuade them with some smoke. Crushing bees helps to spread disease. Care keeps accidents to a minimum.

If you feel that you have missed something, the inspection process can be reversed, one frame at a time. Try to keep the frames left in the box firmly pushed together so that when you get back to the beginning you have the same space left as that with which you started. Then you can gently replace the first frame. Try to keep the same adjacent faces together at all times. Combs are always built parallel but the surfaces may not be exactly flat and level, so putting in a frame back to front will cause problems. If the bees are really filling the box and are building bits of comb on the top bars, use the hive tool to clean these off so that you can put a queen excluder in place, followed by a super.

Reversing a frame

If the bees are bringing in nectar, it may drop from the cells. Firstly, keep the comb over the hive. Secondly, keep the comb vertical using the following technique:

1 Hold the frame by the lugs.
2 Drop one hand down and raise the other so that the top bar is vertical, with the comb sticking out to one side.

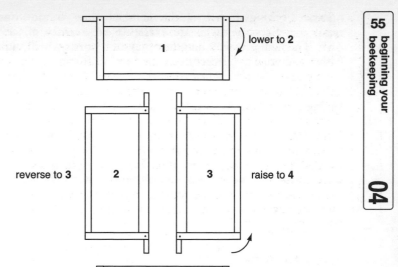

figure 17 reversing a frame

3 In this position, rotate the top bar so that the other side of the comb comes round to face you.
4 Raise the hand you dropped and drop the hand you raised.
5 The top bar will now be horizontal again but the comb is above it (bottom bars uppermost) and you are looking at the other side. The comb has remained vertical throughout. If you are not sure, practise first with an empty comb.

Reverse the actions to look at the first side again or to replace the frame in the hive. Put the inner cover back with the same side downwards. You should be able to drive the bees down from the top of the frames by using smoke but it helps to avoid crushing bees if you put it on first at an angle and then rotate it gently into position. Gently and positively is the way to handle bees.

Then the roof goes back on. However, your inspection is not finished yet. You need to record what you have seen and done in some way. At first, you will probably want to record a number of things to help you remember what you did. Various ready-made record cards are available or you can make your own. This will give you a consistency of records which makes it easier to compare different colonies and different seasons.

What to look for

What are you looking at? Bees can be apparently crawling everywhere – and buzzing. Buzz notes can change and you often hear people say 'Don't they sound angry?'. Certainly, the pitch of the buzz of a bee happily going about its normal business and that of one preparing to sting in the defence of its nest are different. You will soon learn that the pitch of the second one rises significantly.

The brood nest

The brood nest is the centre of the colony. As you now know, brood is the collective term beekeepers use to denote eggs, larvae and pupae. The brood nest is the area in the colony, below the queen excluder, where these juvenile stages are reared and where you will find the queen. Eggs take three days to hatch, larvae take $5^1/2$–6 days to develop and pupae take 12 more days before they emerge as adults. This means that if the brood nest area in the spring is expanding, there should be at least twice as many larvae as there are eggs and twice as many pupae in their sealed cells as there are larvae. As the colony expands, you should be able to record an increasing number of combs containing brood. As I have indicated before, the above timings are correct for worker bees. Remember, drones take 24 days to fully develop while the queen takes only 16 days and workers take 21 days.

If you can think in 3-D, imagine a very roughly spherical brood nest enclosed in a shell of pollen. Above and around this is stored honey. From April to July, if the boxes on the hive seem to be full of bees, then you should add another to avoid crowding. This, of course, includes adding additional supers. If you intend to keep only one or two colonies, you will need four supers per colony.

You will see that the cappings on honey storage cells are thin and vary in colour from brownish on old brood comb to snow white on new honey comb. If the comb in a super is half to three-quarters capped over, or if a similar proportion of combs in a super are capped, then that comb or super (honey box) may be removed.

The brood pattern

The brood pattern should be obvious. You know that this consists of concentric rings, semi-rings or arcs of larvae, eggs or sealed brood. The more obvious it is, the higher the percentage of eggs the queen is laying are viable and the more disease-free your brood is likely to be (see Plate 7).

The primary thing that any beekeeper must see or be able to see is eggs. Eggs confirm that the queen is present and laying. There is no need to see her until the colony is trying to swarm, although as a beginner this can be both reassuring and exciting. The queen's presence is confirmed by regular areas of larvae of different sizes and larger areas of sealed worker brood. Another important thing to note is the food available throughout the brood box. Bees can starve at any period in the year. A full National brood comb holds 2–2.5 kg of honey. The minimum quantity of stores a colony needs is 4.5 kg (the equivalent of two full brood combs). This really is not very much so reserves should never be allowed to drop below this level. I mentioned that boxes full of bees indicate the need for more room. You should never leave a hive in the spring and summer needing more room without supplying it in some way. Expanding nucleus colonies should never run out of food.

Temper

Just about the only thing that non-beekeepers know about bees is that they sting. Bad temper in bees obviously includes stinging. Bees react to rapid movements and contrasts in shade and colour. Unprotected people can be stung near their eyes, for example. Woolly or furry garments or hair can elicit a burrowing response, just as the fur of a predator such as a bear would. Bad bees can also 'follow' the beekeeper, or anyone else for many yards. Some bees bear bad grudges! On the combs, bees can react to rapid, jerky hand movements by jumping and stinging, and some bees run about on the comb making it impossible to find the queen or see what is going on in the brood nest. If your bees' behaviour bothers you, then good-tempered bees can be yours if you requeen the colony. As you become a better handler, your bees are likely to become better behaved.

The right equipment, as described earlier, can make you fairly safe from all this, but your neighbours may not be so lucky. Clumsy handling by the beekeeper can also be responsible for problems with neighbours, but if you, the beekeeper, cannot walk about among your hives without protection, not having touched the bees or opened a colony, then I would say they need improving. Not all bees are bad. You, as a new beekeeper, will become knowledgeable and competent at the craft much faster if you keep bees that are easily manageable.

Swarming

Bees exist as colonies in nature and to make more colonies, bees divide by swarming – the colony divides itself. One of the arts of beekeeping is to control this phenomenon, basically so that the colony does not split and therefore stands a better chance of gathering you a good honey crop.

Inspecting colonies of bees on a regular basis should enable you to detect the early signs of swarming preparation and take appropriate steps for swarm control.

When you inspect the brood nest, you are looking for:

1 Queenrightness, i.e. eggs and a normal brood nest pattern to indicate her presence. The first main thing to note is the presence of eggs in worker cells.
2 At least the minimum of necessary food reserves.
3 Adequate room – after each inspection during April to mid-July, you should leave the colony with a 'super' of space in which the bees can store honey.
4 Signs of swarming preparations.

The natural sequence of events prior to swarming

When bees build up in the spring, the winter population starts to die off, to be replaced by far greater numbers of new bees. Queens produce queen substance and this affects the way worker bees behave. The first obvious change is that bees prepare drone cells in which the queen can lay and, in one of your inspections early in the season, you should be able to see eggs or larvae in the larger drone cells. There may well be pupae as well. Drone cells are larger than worker cells. They are, however, built in the normal comb. The cells can be a little longer than worker cells but, in spite of this, the capping over drone cells is obviously domed. Spacing frames at 35 mm is supposed to cut down production of drone cells. However, I do not think you should do this. To speak in human terms, bees like drones and I think the colony works better with some in it. Drone production can be viewed as no more than a sign that the colony is prosperous, since strong, non-swarming colonies also produce drones. However, colonies capable of swarming always have drones present.

The next step towards swarming you are likely to see is the production of what are known as 'play cells' or 'queen cell cups'. These are produced freely by all strong colonies but they become significant when the workers prepare them so that the queen can lay eggs in them. The first major step you are likely to see is eggs or small larvae in these play cells. Play cells look very like the rough cups that acorns sit in. They have an entrance hole slightly larger than that of a worker cell and rounded walls so that the interior is wider. They are not as deep as worker cells as they are extended by the workers as the larvae grows. They can be very numerous and tend to be constructed around the edges of the brood nest – along the bottom edge is a favourite place (see Plate 15).

If you use two boxes for the brood nest, then play cups and/or queen cells can be found along the bottom edge of the frames in the top box. You can make a quick preliminary search for queen cells simply by splitting the boxes apart and looking along the bottom bars. A full-sized queen cell is about the length of the last two joints of your little finger. They hang down vertically and the larva is held in place by very large quantities of royal jelly. This looks like sweetened condensed milk but doesn't taste as nice! As soon as the larva reaches the right stage in its development, the cell is sealed and it pupates. At this point, the swarm generally leaves the colony. Queen cells ready to be sealed are large and they should be fairly noticeable. Their orientation is different from that of any other cell. They hang downwards and can be 30 mm long (see Plate 7).

Timing of examinations

The best way for you, the new beekeeper, to find out what the bees are doing is to inspect the broodnest regularly. You need to remember the timetable for each stage in the development of the queen:

egg	3 days
larva	5 1/2–6 days (the queen cell is sealed on the ninth day after the egg is laid)
pupa	7 days
total	16 days

If you add the days as an egg to the days as a larva, you get a total of nine. Now you will understand the recommendation that you inspect your colonies at least every nine days. If there really are no signs of swarming on one inspection, even though the bees might 'start' as you leave the apiary, no swarm should actually issue before your next inspection. However, our lives are divided into seven-day periods, so most beekeepers' nine-day inspections are seven days apart. This should be better! I urge new beekeepers to be careful and inspect all the play cells that they can see. Move bees by touching them gently with your gloved fingers or use a whiff of smoke. If you cannot see into a play cell, then open it with your hive tool and make sure you know whether or not it contains an egg or larva.

Consider this. What does it mean if you make a Saturday inspection and miss a day-old larva in a queen cell? Think of the timings. A day-old larva means that you are four days into the nine-day cycle. Five days later, on the following Thursday – two days before you next intend to look into your colony – you will have a sealed queen cell and a swarm. If you missed a full-term egg ready to hatch in a queen cup rather than a day-old larva, the resulting swarm could emerge the following Friday.

Beekeepers shrug and say 'The bees don't read the books'. However, you could say that we, the beekeepers, aren't reading the book the bees have written. Believe me, inspection is the best way. Inspect regularly and carefully.

Seeing the queen

For most examinations, finding the queen is not important. Her presence is indicated by a normal broodnest and eggs in worker cells. Look at the brood. Try not to be distracted by the bees. Look for play cells around the broodnest, especially at the bottom edges of the frame or where the comb has developed holes, for example, and check whether they are occupied. If you are using two brood boxes and the broodnest extends over both, you are very likely to find play cells or queen cells on the bottom edges of the frames in the top box. With experience, examinations can be reduced to removing the supers, splitting the two brood boxes apart and tilting back the top one. Clear the bees from the bottom bars with a little smoke and this gives an excellent chance of spotting occupied queen cells containing larvae.

Swarming is possible at any time between April and September. Some colonies will not swarm at all. The strongest colonies are likely to swarm first and late developers can try at any time. Most swarming is finished by mid-July, but this will be later in more northerly areas. Most swarming takes place between the end of April and the end of June. A successful series of swarm control manipulations means a likely end to the need to examine that particular colony for signs of swarm preparations.

Occasionally, you hear of swarms at odd times such as October and November. In the tropics, bee colonies abscond, with the whole colony deserting its home and going to pastures new. I think so-called 'hunger swarming' is, in reality, absconding. I have collected and fed such swarms and watched them die. There is always something wrong. However, in mild autumns such as 2005, it could just be bees thinking that it is summer again.

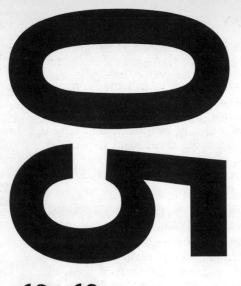

05

swarming and swarm control

In this chapter you will learn:
- simple, workable methods
- if you cannot find the queen…
- after a swarm is lost…
- how to collect swarms
- about hiving and using swarms
- about moving bees.

Swarm prevention

Before you try swarm control, you should be practising swarm prevention. Any colony that survives the winter well and is given no more room than its basic brood box will almost certainly swarm because it will become overcrowded. Beehives allow the beekeeper to add more room as the bees require it. This is why you should have the means to give every one of your colonies three or four supers. You can add empty supers and take away a full one without making the bees feel crowded. Successful beekeeping demands adequate or spare equipment. Without it, you can do nothing. With it, all problems can be solved. Swarm prevention means giving room.

Swarm control

Before your first colony tries to swarm, you must have a plan to deal with this situation and you must also have the equipment ready and waiting. Even if you have nothing ready, depending on what you find, there may well be time to do something. For example, if all the occupied queen cell cups contain only eggs, assume that the oldest is ready to hatch into a larva. That means you have five to six days before a swarm issues. There is time, as I was told once, to go back in the house, have a cup of tea – and think! If you find sealed queen cells, then the swarm has probably gone. I will deal with this situation later.

Any swarm control method involves some form of separation of the parts of the colony. In fact, all effective methods do because they are dealing with the same natural process. Basically, you should think of a colony as if it is made up of three parts: (1) the queen, (2) the brood and its nurse bees and (3) the flying bees. Swarm control involves separating one of these three parts from the other two.

The nucleus method

This was the first method I was taught. It can work very well provided you can find the queen and don't miss any queen cells. In the first step, you separate the queen from the flying bees and the brood.

You will need a nucleus box full of frames, or a spare floor, inner cover and brood box with four or five frames. Have this equipment ready for when you find queen cells.

Let us suppose that during a regular examination of your colony, you find queen cells but none of them are sealed.

1 Put one frame of food (probably an end-comb) plus its bees into the nucleus box.

2 Find the queen and put her and the frame she is on into the nucleus. Remove any queen cells on this frame. This frame should contain a patch of brood about the size of a hand.

3 Put in another comb from the other end of the brood box, i.e. make sure the nucleus has lots of food.

4 Shake or brush all the bees from two more frames into the nucleus. Fill up the box with frames and stuff the entrance tightly with grass. Put the inner cover and roof onto the nucleus hive.

5 Check the frames in the original brood box carefully. Do not remove any queen cells unless they are very large and nearly sealed. Stick drawing pins into the top bars of frames above where you can see tiny larvae in queen cells. This will help you to find them the next week.

6 Push the frames together and fill up the gap with frames you brought with you. There should be enough if both boxes were full of frames when you started.

7 Put the nucleus to one side or on another stand. Allow the grass plug in the entrance to wither and release the bees. If this hasn't happened within 48 hours, remove it yourself at dusk.

The purpose of this nucleus is to keep the old queen alive so that you can use her to replace your new queen if she fails to mate or gets lost on her mating flight.

All the flying bees will leave the nucleus, which is why you shook in extra bees when you made it up. The grass plug system helps make sure that more of the bees you have transferred stay with the nucleus. Don't put too large a patch of brood in the nucleus – you only need one the size of a hand.

One week later

Worker bees can rear new queens from worker larvae that are less than four days old. After this it is not possible for them to do so. On your visit to the apiary one week after making up the nucleus, you should be able to go through the original swarming colony and remove all queen cells except one. This means that there will be no casts from the colony when the virgin emerges.

There should be no larvae left in the colony that are young enough to be turned into a queen. However, check through the colony before removing any queen cells. You never know, the first one you come across (and remove) could be the only one left!

1 First check all the combs you marked with drawing pins. Brush the bees off so that you can see all the parts of the comb. Don't shake them off at this stage. Shaking a queen pupa in a queen cell doesn't do her any good! Choose one cell that you know was occupied the week before.

2 Make sure this queen cell is the only one on that frame. Return the frame to the brood box very carefully. You don't want to damage your new queen before she is born.

3 Examine all the other frames, one by one. Shake or brush all the bees off to make sure that you have removed all the queen cells. Remove anything that looks remotely like a queen cell.

This second visit can be made on day eight or day nine, as long as it is before any queen cells hatch. This makes it even more certain that no more queens can be reared. *Never leave more than one sealed queen cell in a colony.*

Leave the colony alone for two to three weeks, after which the new queen should be laying. This colony should not try to swarm again that year. The nucleus with your old queen can be allowed to build up into another colony if you wish, or united to another colony later.

The artificial swarm

This technique is another two-stage process. One is during the egg/larva period of queen development. The other is just before the queen cells hatch.

The equipment you need is:

- a floor, brood box, inner cover and roof;
- combs or frames of foundation to fill the brood box.

It is helpful to have this equipment nearby when you examine your bees during the swarming season. When you first spot a queen cell cup containing an egg or small larva, I recommend that you start the swarm control process straight away.

An artificial swarm when the queen is found

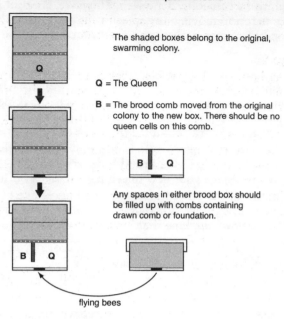

The shaded boxes belong to the original, swarming colony.

Q = The Queen

B = The brood comb moved from the original colony to the new box. There should be no queen cells on this comb.

Any spaces in either brood box should be filled up with combs containing drawn comb or foundation.

flying bees

figure 18 the artificial swarm method

Stage 1

1 Remove all the supers and stand them on the upturned roof.
2 Move the brood box and its floor to one side within easy reach.
3 Place the new floor and brood box in its place and remove three frames from the centre, or at the front if you have your frames the warm way.
4 Find the queen and put her and the frame she is on into the new box. *Make sure there are no queen cells on this frame.* If you have spare drawn combs, place them either side of this comb. Fill the rest of the box with frames of drawn comb or foundation.
5 Put the excluder on top of this brood box and then add the supers, and their bees. Complete the hive with the inner cover and roof.
6 Position the old box to one side, say about a metre away. Face the entrance the same way as it was previously.

7 Examine the frames. Leave all the queen cells you see except for any large ones which you can remove. Push all the frames together and fill up the gap with the spare frames from the first box. Reassemble the hive as before.

Stage 2

Inspect both boxes again one week later. In the moved box (the original colony), there should be several sealed queen cells. When these start to hatch, there is a danger that the first one or two virgin queens that emerge could lead out an afterswarm or cast. What stops a colony producing casts is the loss of its flying bees or the presence of only one queen cell. To achieve this, move this box either to the other side of the queenright colony or to somewhere else in the apiary. The flying bees which have developed during the previous week and learned the position of the new box will return and enter the nearest hive. This will be the one on their original site, which contains the original queen, with all the supers. You will be strengthening this colony with bees. With few or no flying bees, there will be no casts from the colony you have moved and the first of the emerging virgin queens will destroy the others. Providing that none of the queen cells were sealed at step 1, none will have emerged a week later. Leave this virgin and her colony alone for three weeks. Opening and disturbing the colony at this stage could result in the queen being lost if she returns from her mating flight while you are in the middle of your inspection. After three weeks, you should find the beginnings of a new broodnest. Early signs that all is well include lots of pollen being taken into the new queen's hive.

At the one-week stage, check the queenright colony on the original site. Sometimes the bees will have started more queen cells. If so, remove these completely. Check the colony again a week later.

You are less likely to have trouble if:

- there are only eggs in queen cells on the first manipulation;
- there are no queen cells of any sort on the comb of brood you transfer with the queen;
- you can put some drawn combs around this transferred comb of brood, the more the better.

Once the new queen has started to lay, say two weeks after she emerges, she can be moved in her brood box to replace the one containing the original queen. Remove the supers and queen excluder and place them to one side (on the upturned roof).

Move the brood box containing the original queen, plus its floor, to one side and stand the box with the new queen in its place. Cover the top with a single sheet of newspaper and put the queen excluder on top before replacing the supers. The bees will eat through the paper and unite peacefully. They would probably do that anyway, but the paper makes it certain. If one sheet of your regular newspaper won't cover the whole of the top of the brood box, stick two pieces together with masking tape or glue.

The old queen and her brood box can now occupy the site formerly occupied by the young queen. The flying bees should be allowed back into both colonies without any trouble. You now have two colonies which can be united back to one later if you wish. If you doubled your number of colonies each year, you would soon be the largest bee farmer in the world! Do check both colonies regularly. From this point, both may well need supering.

In Stage 1, you separate the brood and nurse bees from the queen and the flying bees. In Stage 2, you separate the new flying bees from the queen cells. An alternative step here would be to leave the hive in one place but remove all the queen cells except one. If you decide to do this, check the queen cells and select one first, then be absolutely certain to remove all the others.

If you want more bees, as a third method you could divide the brood frames between two boxes, making sure each contains a queen cell, and then move them elsewhere in the apiary. You may need a little experienced help if you decide to do this. I will tell you how later.

Cannot find the queen?

Let us say you find occupied and, hopefully, unsealed queen cells but the queen proves elusive. Then I suggest that you repeat the moves of the artificial swarm just as if you had found her.

1 Transfer one comb of brood in all stages – eggs, larvae and pupae – plus bees but no queen cells into the 'new' brood box.
2 Check through the entire remaining broodnest in the original colony and make sure that all queen cells are unsealed. This is important.

3 Put the 'new' brood box, full of frames including the single frame of brood in all stages (step 1), on the original site. Place the excluder on top, followed by the supers and all their bees.
4 Place the original brood box on the new floor to one side, at least a metre away from the new box.
5 Make sure you replace all inner covers and roofs.

Let us look at what you have done. If the queen was on the single transferred comb, you have actually performed the artificial swarm and have separated the queen and the flying bees from the brood.

However, the most likely place to find the queen is in her old brood box on the new site. In this second scenario, you have separated the queen and the brood from the flying bees. The queen is with the queen cells but, since the flying bees return to the new box on the original site, there will be no swarm. Swarms can only consist of bees able to fly and the queen is now accompanied only by younger bees. This should mean that the bees tear down the queen cells in the original box during the next seven to nine days.

At your next inspection, the box with the queen should contain no queen cells while the one without her will have some. Comb in the box containing the queen will have eggs in worker cells.

Examine the queen cells on the frame you transferred. If any are small, remove them and leave the largest one to emerge as your new young queen. You can swap the boxes over as for the artificial swarm, if you wish. However, colonies treated this way quite often don't build queen cells again that year. I would suggest that you carry out normal, regular checks and if queen cells are started again, by then you should have a young, newly mated queen. Swap the brood boxes at this stage and, for one hundred per cent safety, unite the supers over newspaper. If the bees do not make further attempts to swarm, you will have to keep both parts supered.

The first time I attempted swarm control, I couldn't find the queen. I looked many times but with no success. In the end, in panic, I removed all the queen cells. It didn't work. I could cite many more swarm control methods here, but all the successful ones involve recognizing the three parts of a colony – namely queen, brood and flying bees – and separating one of them from the other two. Replace the rule of thumb method by the rule of three. *Never rely on just removing queen cells*!

Losing a swarm

If your efforts to prevent a colony swarming fail, what is the best thing to do? The first step is to make sure you lose nothing else.

Firstly, what makes you think your bees have swarmed? Look out for the following clues:

- The presence of more than just one or two sealed queen cells – there will be only a few eggs in worker cells and, particularly if there are none at all, that confirms that the queen (and the swarm) has gone.
- The bees are often more bad-tempered without their queen.
- On inspection you find all the queen cells are sealed. That means the swarm left days ago. You need to reduce the queen cells to just one large one.
- There are both sealed and unsealed queen cells – remove all the sealed ones and return one week later to reduce the remaining queen cells to just one.

Don't shake the frame which contains your chosen queen cell or you may damage the developing pupa. Just brush the bees off the comb so that you can see if there are any others on the frame. If there are, remove them. You can shake the bees off all the other combs to get a clear view of the whole surface in order to make sure that you have removed all the queen cells.

To shake bees off a comb, hold it firmly by the lugs. If you have a gap in the frames in the brood box, hold it down inside. If not, hold it over the brood box. Lift it up slightly and then jerk down sharply, not forgetting the sudden stop which will shake the bees off. You may have to repeat the action. You have to remove only enough bees to see what you want to see.

It may well be that the swarm left a week prior to your inspection of the colony. You might find one or more queen cells open at the tip – with no occupant. In other words, the bees have started to produce casts as the virgin queens have hatched. The thing to do in this situation is to examine all the queen cells. Some will be ripe and the wax will have been removed from their tips revealing the brown parchment-like cocoon. This means that a virgin queen is ready to emerge, maybe quite soon. There may also be a virgin queen already running loose in the colony.

Using the corner of your hive tool or a penknife, start to open the tip of the queen cell. Sooner or later you will find one with a living queen inside. Remove the tip and release her. Then you must remove all the other queen cells. Surprisingly, you can actually release more virgin queens into the colony if you wish. What makes a colony produce casts is plenty of bees, a free-running virgin queen and occupied queen cells. If there are no queen cells, there will be no casts. The other things that will prevent casts are the absence of flying bees or the fact that the colony is weak.

Don't leave two queen cells just in case. Do release one young queen and you should be fine. Check the brood nest two weeks later to make sure that your virgin has mated and is beginning to lay. Then your colony is back to normal.

Collecting swarms and hiving them

Once your non-beekeeping friends know you keep bees, you may well be called on to collect a swarm (see Plate 16). The swarm may well be your own! The two main parts of queen substance come into operation during swarming. One part keeps the bees with the queen when they are flying and the other keeps the swarm stable when it clusters.

Swarms are quite easy to collect if they are free hanging at about chest height. More often than not, they are at the tips of branches our motroo in the air with no easy way of getting at them. For these high swarms, you may find that a swarm catcher helps. This is a large bag which you mount on a long pole. It has a metal rim which can be closed by pulling on a cord, so you may well be able to collect a swarm 4–5 metres off the ground with little trouble.

Bees inside chimneys or cavity walls are very difficult to deal with and I would recommend that, as a beginner, you either enlist help from a more experienced beekeeper or pass the call on to somebody else.

What will you find when you go to collect a swarm? If it is a prime swarm, you will see a mass of bees, roughly the size of a football but elongated. I think the largest swarms come from single colonies weigh about 3 kg. Anything larger than that comes from two or more sources. I have certainly collected a prime swarm of only 1.5 kg. A prime swarm is one where the queen is already mated, i.e. the first swarm from a colony. A cast will contain at least one unmated queen.

The equipment you are likely to need is as follows:

- a pair of secateurs or loppers
- a receptacle for the swarm – this can be a straw skep, a nucleus box with a floor, a Modified National brood box or similar with a floor, or a cardboard box
- stout string or a hive strap
- your smoker, smoker fuel and matches or a lighter
- a large open-weave sheet/net curtain which is big enough to cover the aperture on your receptacle with plenty overhanging so that you can tie it on securely
- your bee veil – in spite of what people say, swarms can be nasty tempered
- access to a step ladder if the swarm is up high.

As long as you get permission from the owner of the tree, the first thing to do is to cut off any intervening branches around the swarm so that you can get at it easily. If the swarm is free hanging, hold the skep (or other receptacle) underneath the cluster and bring it up as high around the bees as you can.

Shake or strike the branch sharply to dislodge the bees into the skep. What you are trying to do is to get the queen and as many bees as possible into the skep. The queen is the key to successful swarm collection. If she is not in the skep, the swarm will not want to stay there and you will find yourself having to repeat the performance. If the branch or whatever the bees are hanging from is thick, get someone else to hit or jerk it while you hold the skep in place. It is more difficult than you think to support a receptacle as several kilos of bees drop into it.

If the swarm is on a solid support such as a gate post or fence, you have two options. You can use a brush, goose feather or your (gloved) hand to dislodge the bees into the upturned skep. Alternatively, if you can secure the skep above the swarm, you can persuade the bees to crawl up into it by gently applying smoke at the bottom of the cluster. Think like a sheep dog!

When you have the bees in the skep, put your cloth over the top to stop the bees flying out. Return to the ground if you have had to climb a ladder to reach the swarm, then gently invert the skep and cloth and put them on the ground near to where the swarm was hanging and spread out the cloth. If your receptacle does not have an entrance, prop it up at one corner to give the bees access. Your nucleus box or hive body should have an entrance which you can open.

plate 1   worker honey bees returning to the hive with large pollen loads

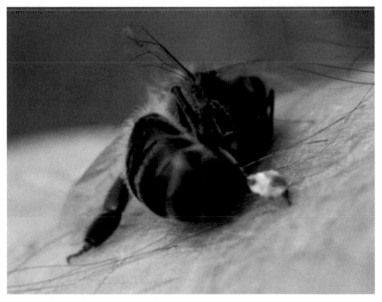

plate 2   the sting

plate 3   new natural comb built in a straw skep

plate 4   the queen honey bee

plate 5 the worker honey bee

plate 6 the drone honey bee

plate 7 sealed queen cells around the edge of the brood nest

plate 8 the domed drone brood cappings stand out above the flatter ones of the worker brood

plate 9 worker honey bees on the comb

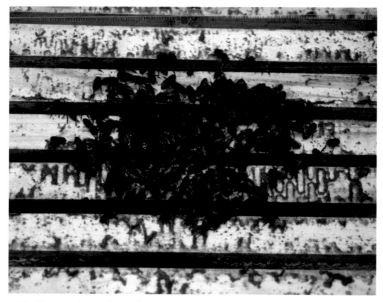

plate 10 the cluster in winter

plate 11 a worker honey bee fanning and exposing her Nasenov gland

plate 12 a protective veil for use during manipulations – note how the hive tool is held in the hand

plate 13 the smoker and hive tool

plate 14 honey bees drinking from wet peat. It is important to keep a water fountain topped up once bees have started to use it

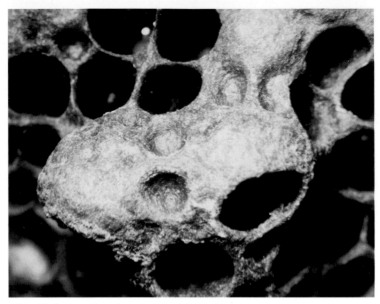

plate 15 'play cells' – how queen cells begin

plate 16 a swarm hanging in a tree

plate 17   hiving a swarm

plate 18   a worker honey bee on pussy willow (*Salix caprea*)

plate 19 a worker honey bee on blackberry (*Rubus fruticosus*)

plate 20 this colony is ready for a super

plate 21   a larva of the greater wax moth (*Galleria mellonella*) in its webbing

plate 22   a contact or bucket feeder

plate 23 pouring sugar syrup into a Miller feeder

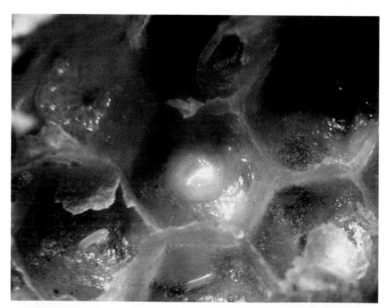

plate 24 a newly hatched larva in a pool of royal jelly, surrounded by eggs and (top left) an older larva – the cell walls have been cut down to show the contents

plate 25 sealed worker brood – note the ample pollen stores

plate 26 larvae killed by the chalk brood fungus (*Ascosphaera apis*)

plate 27 European Foul Brood (*Melissococcus plutonius*) – note the abnormal positions of the larvae in their cells

plate 28 the roping test for American Foul Brood *(Paenibacillus larvae subsp. larvae)* – note the sunken, greasy cappings on the sealed brood

plate 29 a worker honey bee with a varroa mite on her abdomen

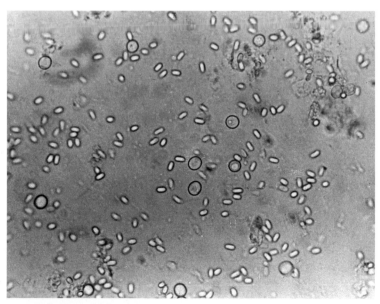

plate 30 the rice grain-shaped spores of *Nosema apis* found in the gut of an infected honey bee – the small circles are the protozan *Malpighamoeba mellificae*

plate 31 an apiary in winter

plate 32 a worker honey bee collects propolis from an old hive – note how she packs it into her pollen basket using her middle leg

Many bees will start to fan around the box, but others will try to return to the clustering site. Use your smoker to drive them away from there and into the air. If you have the queen in the mass of bees in the skep, her presence will make the others want to join her. They were looking for a suitable cavity which you have very generously provided!

Soon you should see more and more bees landing around your skep and going inside. All you have to do then is to wait for the flying bees to join the rest of the swarm. You may have to leave things there and return in the evening when the bees have stopped flying for the day. If you are happy that you have, in fact, got most or all of the swarm into the receptacle, take away the prop or close the hive entrance. Secure the sheet around the container and make sure there are no small gaps through which bees can escape. Then turn the whole thing over carefully so that the bees can breathe through the cloth.

You have collected your first swarm!

Swarms know where they have come from but choose to forget it as their purpose was to move to a new home. This means that you can hive the swarm anywhere you like in your apiary, even if it has come from your own bees. If you leave any bees at the clustering site, these will return to their original colony.

The best use of a swarm

If the swarm is indeed yours and you know from which hive it came (this is easy if you have only one colony!), I recommend the following course of action. You will need a flat board that is about the width of the hive and will reach from the ground to the entrance (see Plate 17).

1 Light your smoker.
2 Smoke the colony that produced the swarm and move it to one side (in pieces – you want the supers).
3 In its place, put a brood box full of frames with foundation.
4 Put the queen excluder on top of the box of foundation and then add the supers and their bees.
5 Put the board in position, touching the hive entrance. If you are using a shallow floor, I suggest that you prop this up to a height of 12–20 mm.
6 Place the swarm, cloth side down, on the board and undo the ties.

7 Spread the cloth out as evenly as possible and make sure that it doesn't block the hive entrance. It helps if the cloth hangs over the edges of the board down to the ground as this stops bees finding their way underneath.

8 Lift the receptacle about 30 cm and shake out the swarm with a firm jerk. Don't let go of the box! One or two more shakes should dislodge the stragglers.

9 The swarm will spread out on the board.

10 Some bees will find the hive entrance and, gradually, the mass will turn and begin to march inside. Don't let them climb up the front of the hive too much. Just brush them back down. I use a goose feather but you can use a handful of long grass or your hand to gently knock them off. Watching a swarm walk into a hive is one of the finest sights in beekeeping. You may even seen the queen making her grand entrance!

11 As soon as you think they are all inside, remove any prop so that everything is level again.

Next day, the swarm will collect the flying bees from the colony that was originally on the site and this strengthens it. You must inspect the colony you moved and remove all of its sealed queen cells and leave all of its unsealed ones. One week later, go into it again and reduce the remaining queen cells to just one. Don't let anyone who should know better tell you to leave two or you will lose a cast when the first virgin hatches. This method was devised by Mr Pagden who, in 1868, was able to write a booklet entitled £70 *a year: How I make it by my bees*. A lot of money in those days. This booklet was very small and is now very, very rare. You would be lucky to find one.

Making increase

It may be that you wish to make an increase in your number of colonies. An increase can be useful on a temporary level and having several new colonies allows you to make a choice as to which ones you wish to keep, which is always an advantage. For the complete newcomer, the basic artificial swarm offers an excellent opportunity to make some increase. If during the summer you create more colonies than you wish, they can be united to create the number of colonies you want in the autumn.

You will remember that the queen was left on the old site with the supers and the original brood box was put to one side for one week. At the time this box is due to be moved again to shed its flying bees to the queenright colony, it can be divided and the divisions put in different positions in the apiary. As a new beekeeper, I would suggest that you should make just two or three divisions. Remember that when you split up the frames in a brood box in this manner, the flying bees will return to their 'home'. You must allow for this so that enough bees stay in each box to keep the brood warm. You must also have sufficient boxes for the divisions you plan to make. These can be nucleus boxes, small hives built in the same pattern as the brood box but holding only three, four or five frames. Most commercially produced nucleus boxes hold five frames. This does not mean that they have to be filled. Initially, for example, a three-comb nucleus of bees will fit into a five-frame box – but not vice versa! You can also use full-sized brood boxes with the one containing the queen cells serving as one.

Dividing the colony

Proceed as follows.

1 Prepare your nucleus or other brood boxes and place them nearby.
2 Open the brood box containing the queen cells. Survey all the combs to assess how many queen cells are present and decide how many divisions you are going to make. I suggest two.
3 Transfer a comb of food (usually found at each end of the broodnest) into each of the appropriate number of boxes.
4 If the broodnest is to be divided into two, put two combs of brood and bees into each box with at least one good queen cell for each nucleus.
5 Brush the bees from any other brood combs into the boxes, distributing them reasonably evenly. Remove any queen cells from these combs, including anything that looks even remotely like a queen cell. Swap these combs for broodless combs in the queenright colony. You are giving it more bees when the brood hatches. Put these combs in or near the broodnest.
6 In the divisions, put the brood combs against the side wall of the nucleus box with the food comb on the outside, or between the food combs if you have two. In full-sized brood boxes, place the combs together in the centre of the box with combs of foundation on each side. You can also use a dummy frame on each side. A dummy frame is just as it sounds – a

frame made from wood rather than with comb. They can be bought, or made by nailing a top bar to a thick piece of chipboard or plywood. The finished dimensions should be size of one of your normal frames.

Choosing queen cells

Place the divisions away from the other hive. Remove only very small queen cells and leave the good ones. These will be slightly tapered, have a rough textured surface and a large base. Just before the queen inside is due to emerge, worker bees remove the wax on the outside of the bottom third and tip of the cell. This makes that part look browner and rather like parchment. Beekeepers call such cells 'ripe' queen cells. The queen can emerge from such a cell at any time. There is no need to remove any queen cells in your divisions. Diverting the flying bees means that when the first queen emerges she will be allowed to damage all other queen cells. Unlike you, she won't miss any!

Do not keep looking into your colonies, nuclei or divisions to see what is happening. You could be disturbing things just as the queen returns from a mating flight and she could enter the wrong hive. The bees there will, of course, kill her. When bees kill queens, they do so by forming a tight ball around her. I don't know if she is stung or just mauled, but she usually dies.

As soon as the young queens mate and begin to lay, you may have to start adding combs or frames of foundation to these divisions, or nuclei. A nucleus is simply a small colony ready to develop into a larger one. Three weeks later the populations will start to grow and you must make sure that the little colonies have combs on which to expand. As soon as the nucleus is full of bees, it must be transferred to a bigger box.

Uniting colonies

Five or six combs, packed with healthy bees and food are likely to be a large enough unit to survive the winter. However, strong colonies are best for wintering. Additionally, your bees will have to be protected against *Varroa destructor*, a parasitic mite which kills the colony if bees are left to their own devices. You need to make a decision on the number of colonies you are prepared to treat and feed for the winter. The nice position to be in is to have too many colonies so that numbers have to be reduced. You can kill the queens in the colonies you like least and unite their bees to a colony you like much better. I wish all beekeepers did that.

Uniting can be achieved in many ways, but the simplest and most effective is to use newspaper. The bees chew through the paper and mingle peacefully. Quite why this works is uncertain but it is probably because this process allows populations to mix gradually. Using your hive tool, clean the wax and propolis from the top bars of the colony containing the queen you wish to keep. Find the queen you wish to depose and kill her. There is no special way of doing this – just squash her, between finger and thumb or foot and ground. Check the bottoms of the frames in the box that will go on top and remove any pieces of protruding comb.

Place a complete sheet of newspaper over the box containing the queenright colony. To be safe, you can fold down the excess and pin it in place with one or two drawing pins or staples. It is surprising how often on an otherwise completely still day a little gust of wind arrives just as you are picking up the other box to place it on top, and the paper blows out of position! You can also make a small hole in the paper if you wish (with the corner of your hive tool, for example) but this is not actually necessary. Place the queenless colony on top of the paper and ensure that the boxes are aligned.

You can combine two full colonies or three or more small ones. The object is to achieve a colony of sufficient wintering size, headed by a good queen producing normal brood.

There are a few 'rules' which I think should help. I know of other beekeepers who do things differently. Of course, I think that I am right!

1 If the colonies to be united are dissimilar in size – the weakest goes on top.
2 If the colonies to be united are the same size – the queenless one goes on top.
3 The colony that you have moved goes on top.

Uniting can take place without the removal of any queen but the beekeeper then has no control over which one will survive. I strongly recommend that you attempt to eliminate the worst behaved of your colonies at the end of each year. You will find that this will result in overall better temper in your bees and you will therefore have even more pleasure keeping them. The greater strength of the united colony means that chances of winter survival are improved and spring development is also better. This means that it will be easier to restore/increase colony numbers through division and you will also be developing more colonies from your best ones.

After uniting, the combs will need rearranging. The next day, after placing the paper between the two boxes you will begin to see newspaper dust being thrown out of the entrance. Leave all alone for a week. Then separate the two boxes and put all the brood into one box. In August or September, it should be possible to do this. Try to put combs together that have similar sized patches of brood. Copy the bees – largest patches in the middle, smallest on the outside. Then do the same with the combs containing honey in the second box – heaviest in the middle, lightest at the outside. Put this box of food on top of the brood. There is often enough honey in the combs from the two colonies to provide sufficient winter feed for the combined colony. In the spring, restore the colony to one brood box or run the combination as a double-brood colony, but don't expect double the honey crop.

Moving bees

Uniting colonies may well require you to move colonies within the apiary. Flying bees will return to the spot where their hive stood. After a while, if they find nothing there, they will enter the nearest colony. If these are within half to one metre, you will see this beginning within minutes. It will take a little longer if hives are obscured from one another by landmarks such as bushes or sheds, or are further apart.

If you must move your colony over a long distance, say greater than 20–40 metres, or there are objects in the way, you can move it in a number of short steps. Bees can re-locate their hive more quickly if it is moved in the direction that it is facing, especially if this is backwards. If the move is across an open space, it can be two metres or so. Forward movement should be around one metre and any move sideways about half a metre. Allow at least one or two days of normal flight between these short moves. The whole move may take days or even weeks. You will help bees 'keep track' of their hive if you place a large white flat object in front of it to serve as an identification mark for the bees. Move it with the hive.

If bees are moved beyond their flight range, they re-orientate to the new position of their hive. If follows, therefore, that a single move to a different site several kilometres away, with a 'holiday stay' of one to two weeks, followed by a return to the new position in the original apiary would solve the problem. Even then, bees have been known to remember their old site and go there. My advice in these circumstances is to leave things in their new positions. If bees are still at the old site at dusk, they will be in a cluster which can be shovelled or brushed onto a suitable receptacle and shaken onto the inner cover of their hive at the new site. Usually, bees returning to find their hive has gone realise their mistake and, although they may make a fuss, they all find the new location.

Moving bees for three kilometres or more can put them at risk if you do not take suitable precautions. Bees breathe and need air. Hive parts are separate and, although having a hive fall apart in a moving car may be a challenging experience, it is not one with which your insurers would sympathize. You can buy straps designed to fasten hives together securely. You can also buy or make mesh screens to fit on top of the brood box of your hive. The hive then undergoes its journey without its roof on, but don't forget to take it with you. It will be needed at the new location! The colder the weather, the less likely it is that you will need extra ventilation. Being confined with inadequate air supplies causes bees to panic. The temperature in the hive rises and, at 40 °C (105 °F), the comb collapses and the colony can die.

When you move a colony, you must block the entrance. I use plastic foam, such as that found in upholstered cushions. There are various fastening devices available from equipment suppliers but probably the easiest is a hive strap. Fasten your hive together and, early in the morning before the bees begin to fly, block up the entrance.

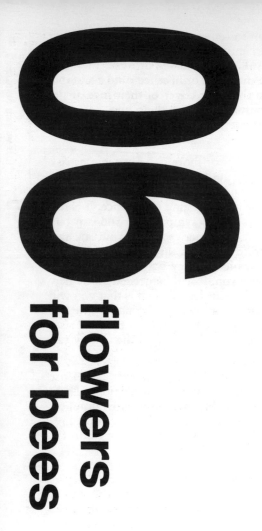

06

flowers for bees

In this chapter you will learn:
- about the relationship between bees and flowers
- what bees see
- how bees 'talk' to each other
- about honey supers and honey storage
- how to take honey and its extraction
- about beeswax.

Pollination

We all know bees visit flowers. Flowers want insects for pollination so they provide two attractants – nectar and pollen. The pollen has to be moved from the stamens to the stigma, i.e. from the male part of the flower to the female part. Most flowers provide lots of pollen, and bees as well as other insects use the surplus as food. Flowers also produce nectar, a sugary, sweet liquid. This is purely there to attract insects, not just bees. However, insects adapted to live on nectar and pollen make the most efficient pollinators. Pollination is also achieved by bats, birds, beetles, flies, mice, the wind and water – the list is very wide ranging.

Nectar is composed of a variety of sugars. It is really bee food but its high water content means that it will not keep, as yeast cells enter the nectar and it starts to ferment. In the United Kingdom, bees reduce this to 17–18 per cent moisture and it will then keep.

Many different plants are visited by bees, (see Plates 18 and 19) but relatively few provide enough nectar for a honey surplus. Bees in towns, especially large towns like London, Birmingham and Manchester, do very well, especially in the spring when early flowers, bulbs and shrubs are plentiful. Winter-flowering shrubs, *Lonicera fragrantissima* for example, smell strongly just to attract bees and, at the time they flower, have little competition.

Bees in flowers

It is interesting to observe honey bees working in flowers. They spend lots of time in some and appear to know exactly where to go to collect their reward. Other flowers on the same plant are paid scant attention. Although bees visit flowers for the nectar and pollen, they can sometimes be seen in the flower actively brushing the pollen from their bodies and rejecting it. Some flowers produce little or no nectar but still manage to attract bees with their pollen.

If they are very hungry, bees will occasionally work nectars which have a concentration of 10–12 per cent sugar, but not otherwise. Flowers on pear trees are very attractive to bees but produce nectar with only about 12 per cent sugar content. In tea or coffee, this would be too syrupy for even the sweetest tooth to enjoy! It is the pollen in the pear flower that attracts the bees.

The protein content of fruit tree pollen is high and it is produced at a time when colonies are growing in size and there is plenty of brood to rear. Bees can rear a small amount of brood without having access to additional pollen by using reserves in their own bodies. Young bees reared like this cannot do the same when they reach the appropriate age because they have no reserves. Nectar will keep adult bees alive but will not allow these bees to rear more.

Garden flowers

You may well alter your gardening practices when you start keeping bees because watching your bees working your flower border makes it come alive in a very special way.

Plants which flower in winter and early spring are often highly scented. *Viburnum bodnantense, Viburnum tinus* and *Lonicera fragrantissima* are all obvious in this way, even to the human nose. Bees visit all of them in the winter and less often when there is more on offer. All spring bulbs – snowdrops, aconites, crocuses, chionodoxa, etc. – have flowers attractive to bees. Daffodils, however, are only worked by bees when nothing else is on offer and, similarly, I have never seen bees working the ever-present Forsythia.

Nectar guides

If you look at the striped crocus flower, you will see the obvious patterning. We humans have developed species where these marks on the petals are emphasized because we find them pleasing to our eyes. Bees and other insects see these as signposts. They are known as nectar guides. Such guides can be present as stripes, patches or blotches of colour. Remember, though, that bees can see ultra-violet. We humans see our rainbow, but bees see their own spectrum (see Figure 19).

Flowers that appear plain to us may well have nectar guides which reflect ultra-violet light and are thus visible to the bees. Detergents contain 'blue whiteners'. These reflect ultra-violet and your whites seem brighter. Bees will see this as a colour. Perhaps this is why they sometimes settle on washing on the clothes line.

Nectar guides can also be present as scent markings which are equally obvious to the bees' smelling organs, their antennae. Bees have a hard outer body covering, their exoskeleton.

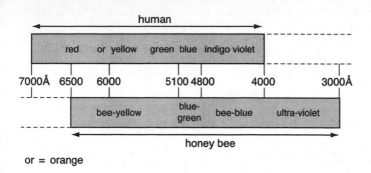

human

red | or | yellow | green | blue | indigo | violet

7000Å 6500 6000 5100 4800 4000 3000Å

bee-yellow | blue-green | bee-blue | ultra-violet

honey bee

or = orange

figure 19 parts of the spectrum seen by the human and the honey bee eye

This has a waxy surface and this waxy cover is very good at picking up flower scents. Other bees can also pick up flower scents from their companions' bodies.

Up to five per cent of the bees going out to forage are there to look for new sources of food and the other ninety-five per cent are there to exploit what the scouts find. In order for this to work, bees have to be able to communicate, and they do. Their language is a combination of scent, sound and movement. Professor Karl Von Frisch studied these movements for most of his working life. We known them as nectar dances (see Taking it further).

The forms of the dance are very similar for all the main races of honey bees but, although the form is the same, the message is not. Therefore, not only do bees have a language, they also have dialects. The Egyptian bee might dance to indicate that food is 10 metres away but the same dance would tell an Austrian bee to fly and search 100 metres away.

There are two main forms of these communication dances – the round dance and the waggle dance. Occasionally you find bees that continue to dance on the comb even after you have lifted it out of the hive into the air. Their movements were noticed even in the eighteenth century, when it was thought that they were dances indicating the bees' joy on sunny days.

In the round dance, the bee runs round in a circle and, on completing the circuit, turns round and runs round the same

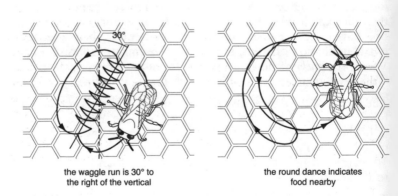

flowers 600 m from the hive,
30° to the right of the sun

30°

600 m

30°

the waggle run is 30° to
the right of the vertical

the round dance indicates
food nearby

figure 20 honey bee communication dances

circle the opposite way. This action of running in circles is repeated again and again. Bees never run continuously round and round but always reverse direction with each circuit. The dancing bee 'recruits' followers, or at least some interested bees. The bee's waxy exterior has picked up scents so the followers know the scent of the flowers for which they should be searching. These round dances are simply saying, 'Go out of the hive and look everywhere within 10 metres'. Bees drinking syrup from feeders perform round dances which is why any syrup spilt on hives or on the ground is so quickly found. Even syrup has a scent so such careless spillage can encourage robbing.

For every round dance I have ever seen, I have seen thousands of waggle dances. There are other dances which indicate food sources between, say, 10 metres and 100 metres, but sources beyond that are indicated by the waggle dance. This indicates direction, distance and the nature of the forage from its scent.

The dancing bee runs in a straight line. As it does so, it vibrates its abdomen rapidly from side to side, maybe at 15 'beats' per second. Try moving your tongue at that speed! The bee finishes the waggle run and turns, say, to the left, and runs round in a rough circle back to the beginning of the line indicated in its waggle run. It runs forwards again along the same line, wagging its abdomen as before. This time, at the top of the run, it turns to the right and returns to the beginning again. These actions are repeated several times. Interested bees touch and follow the dancer. They pick up the scent of the discovered food source from the body of the dancing bee. From the vigour of the dance, they get an assessment of how rich the food source is and they learn how far it is from the hive by the number of waggling runs per 15 seconds. The fewer the number of waggle runs, the further the food source is from the hive. The following bees make tiny, virtually inaudible buzzing sounds. On detecting these, the dancer stops and feeds its followers with some of the nectar it has collected from the flowers it has found.

The direction of the food source relative to the hive entrance is indicated by the line of the waggling run. All this dancing takes place in the darkness of the hive, so a point of reference is needed. The bees use the sun and make allowance for the fact that the sun moves across the sky during the day. As long as they can see a bit of blue sky, bees know the position of the sun.

Inside the hive, the direction 'towards the sun' is vertically upwards. So, if the direction of the food source is towards the sun, the waggle line is straight up the face of the comb. If the source is 30° to the right of the sun, the waggle line is 30° to the right of the vertical, and so on. If further dances indicating the same source occur through the day, the dance will change direction to allow for the sun's movement. The angle is always that relative to the position of the sun.

There is also a magnetic element to the dance. Bees are aware of earth's magnetic field and Von Frisch found that if he cancelled out this magnetic field by placing an equivalent magnet around the hive, the dances were disoriented. In places in the tropics where the sun actually passes directly overhead, the dances are disoriented for the brief period at noon when the sun is vertically above the colony. Not only is this dance language remarkable but so is the work of Karl Von Frisch. Together with Konrad Lorenz and Nikolaas Tingbergen, he was suitably rewarded with the Nobel Prize in 1973.

Bees will actually visit anything that rewards them suitably for their efforts. Aphids feed on plant sap but reject the sugar in the liquid they suck in. This sugar falls on leaves and, after being semi-diluted by dew, it is collected by bees and other insects. It is known as honeydew. Sap-sucking insects on the pine trees of Germany produce *Tannenhonig*, or pine honey, sometimes in large quantities.

Leaves can also produce nectar from spots called extra-floral (i.e. not in the flower) nectaries (nectar-producing places). The most common example in Britain is the cherry laurel (*Prunus laurocerasus*). Occasionally, when the leaves are young, honey bees can be seen working these extra-floral nectaries avidly.

If you want to label your honey as coming from a single nectar source, such as clover or acacia, you have to be able to demonstrate that what you claim is true. Single-source honey can only really come from places where bees have next to nothing else to work, such as when they are on the heather moors.

Honey flows

As a new beekeeper, you will hear others talk about the 'honey flow'. Bees get some sort of income from nectar and pollen almost any time when they can fly and leave the hive. In the spring, what a colony can bring in as nectar is likely to be insufficient to meet the colony's needs. Most of the food fed to bees in the autumn is consumed in the spring to help colony development. If you did not feed enough before winter, then you should feed in the spring. A colony needs the food to help create new bees. It needs the space created by eating the food to provide the cells in which to rear them. If bees really fill up all the potential spaces in a hive (with themselves) then be prepared to give extra room at any time. Modern varieties of oilseed rape (canola) can be seen in flower at the end of March. I have known beekeepers who supered strong colonies then and, in spite of the earliness, had them occupied. Any beekeeping instruction you receive should always be interpreted with common sense.

There are several patterns to honey flows in Britain. For those who keep bees in towns, I am told that some sort of honey flow is possible between April and October. Many years, there will be surplus in this period from, for example, the lime trees (*Tilia* sp.).

In the United Kingdom there are normally two periods when honey flows can be expected. The first is between April and May, followed by a period of relative dearth in June. This is followed by the 'main' flow from late June to early August. The first part feeds the build-up to the swarming period and the main flow allows the swarm to feed itself for the winter. Some areas distinguished by altitude or northern latitudes have a season which begins late and goes through until the main flow when the heather flowers in August. Bees in such areas often swarm in July. Even with a relatively small country like Britain, there are great climatic variations which have their effect on bees. For example, those in Cornwall may fly on many occasions in the winter, while those in the north-east of Scotland may well be confined to the hive by cold for many weeks. Some areas have strong early honey flows but very little after midsummer. In the past, there were undoubtedly slight variations in the native bee which allowed bees to flourish in most places. Bees are not all the same. The various climates in the 'old world' have produced a large number of bee 'races'. All these were adapted to their local climatic conditions. Many, like our native bee, have been 'damaged' by cross-breeding with imports. I much prefer native bees which are better adapted to our conditions.

What makes honey?

Honey flows require hot weather, but not necessarily sunny weather. High humidity as well as heat seems to enhance nectar production in any flower. Very hot, dry weather is not so good. The sap in the plants dries up and therefore there is no nectar.

What you will see at the hive will tell you what is happening. Bees work at a high rate. High rewards make them hurry to unload and return to the field for more loads. The returning bees have distended abdomens. They land heavily. As they manoeuvre to land, they often seem to have their legs forward, presumably to counterbalance the weight of their abdomen. Around or near the entrance, particularly in the evening, other workers can be observed fanning. They face in towards the entrance, their abdomens curved slightly downwards, and beat their wings vigorously. There can be many of them if the honey flow is good and the roar of their fanning can be heard quite a distance away. The bees are fanning away from the entrance and, in so doing, drawing warm, moist air out of the hive. This has the effect of 'ripening' the nectar. To store the nectar that the bees collect as their food requires the water content to be reduced to the point where it cannot ferment. Nectar in the United Kingdom has an average sugar content of 30–40 per cent. For it to keep, honey must have a water content of 18 per cent or less. It is illegal to sell honey with a moisture content of much more than 21 per cent.

If honey flow conditions are good and the colony of bees is strong, such a colony can easily put on 4.5 kg in weight in a day. It is quite possible for bees to fill a super of already built (drawn out) combs in three days. It is also possible for a colony to starve to death in summer, so don't take getting a honey crop for granted.

Bee plants

These are just some of the plants bees visit to collect nectar and/or pollen. Many more could be added.

Aconite

Aubretia

Blackberry

Borage

Buddleia globosa

Butterbur

Cherry

Chickweed

Chinodoxa

Clover

Crocus
Dandelion
Field Beans
Field Maple
Fuchsia
Gooseberry
Ground Ivy
Hazel
Heather (Bell)
Heather (Ling)
Hebe
Helenium
Horse Chestnut

Ivy
Lime
Michaelmas Daisy
Norway Maple
Oilseed Rape
Raspberry
Shrubby Lonicera
Snowdrop (single varieties)
Sweet Chestnut
Sycamore
Willow
Willowherb

Supering

A super is a box which is placed on a colony as a space in which the bees can store their honey. The queen is prevented from laying eggs in it by placing the queen excluder underneath. To take advantage of the instinct bees have to extend the cells in which honey is stored, super frames can be spaced more widely than those in the brood box. The widest spacing in British Standard hives is achieved by using 'wide ends'. These are spacers made of either plastic or metal, which fit onto the frame lugs and then butt up to each other, preventing the frames from moving closer together. I have extracted 55 kg of honey from 32 wide-spaced frames on a WBC hive. These were in four supers, i.e. eight in a super which would normally take ten frames. Frames of foundation spaced at this width are too far apart and the bees will build comb both on and between them. The result is a real mess. To use this wide spacing, you need to have frames containing drawn comb.

A good compromise is to use ten frames in a Modified National super. These can be ordinary SN1 frames spaced with ten-slot castellations (notched metal runners). You can also use ten wide-ended 'Manley' frames per super. Bees will build proper comb on frames that are so spaced. The slightly fatter comb makes them easier to uncap.

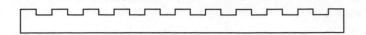

figure 21 a ten-slot castellation

When to super

There are no hard and fast rules to guide you in deciding when to put your supers on. Use your common sense. I think many beekeepers start looking inside their colonies too soon in the year. The right time is around when the flowering currant (*Ribes sanguineum*) is in flower. The first super goes on when the brood box (or boxes) are full of bees. You will have to take the inner cover off to be able to see if this is the case (see Plate 20).

First clean off the top bars in the brood box using the hive tool. Then put the queen excluder in place. You can do this after a regular brood nest inspection. Contrary to many, I prefer the flat metal excluder (see Figure 10). Older versions were made from zinc but the modern ones are of galvanized steel and much stronger. You can protect it with a thin layer of petroleum jelly which will help to stop the bees sticking it down with propolis. It is easy to say 'brush with petroleum jelly' but it takes far longer to actually do this.

The other type of excluder available is made from spaced wires, framed in wood. Hives with a top bee space – above the frames – need a rigid excluder and you therefore have to use a wire excluder in these cases. Older versions had the wires centred in the wooden frame but modern ones have the wires level with one surface, with the complete bee space below or above, depending on how you put it on the hive. Because the top bee space Langstroth hive is oblong, the wires in the excluder have to go whichever way they have been set by the manufacturer. With hives using the British Standard frame, the boxes are square and you have a choice.

I think all queen excluders spoil horizontal bee spaces in hives and I doubt the modern version of the wire excluders are very much better. All the ones I have seen after a season's use, whether wire or flat sheet, are propolised and have brace comb built on and through them. Having said all that, I would not like to keep bees without using queen excluders.

Smoke the bees down gently and lay the excluder with the slots at right angles to the frames. However, if you are using a wire excluder, the wires should be placed parellel to the frames. When the first super is full of *bees*, not necessarily full of honey, a second should be given. You will have noticed that the brood nest has a centre, where the patches of brood on each comb are largest. The bees are likely to first enter the super above this area. If you have one or two 'drawn' super combs, this is the place to put them. Do not alternate foundation with drawn comb as you may be advised. You will get fat comb built on the drawn comb and thin comb on the foundation. It is better to put one drawn comb over the centre of the brood nest and two others at the ends of the super.

Remember, I advise ten frames in an eleven-frame box, using metal castellations to space them evenly. Castellations are thin strips of metal which are attached to the side walls of the super. They have slots which take the frame lugs and they are attached so that the tops of the frames are level with the top of the box (for bottom bee-space hives) or so that the bottom bars are level with the bottom of the box (for top bee-space hives). They come with different numbers of slots to accommodate different frame spacings and, in this case, I would be using ten-slot castellations in my supers. Don't use castellations in the brood box as you need to be able to slide the frames up against each other. There are metal runners available for use here.

As mentioned, the second super goes on when the first is full of bees. You may be sticking rigidly to your 7–9 day inspection routine as I have been advised. In good honey flow conditions, I have known bees fill a super in less than a week. Visits at weekly intervals may mean the bees could be short of room before you return. The answer is simple. Super in advance of requirements. In summer time, an extra super causes far fewer problems than insufficient supers. Lack of super room means that honey is stored in the brood nest which makes the bees feel congested and is likely to push them towards swarming. You can put on two supers at once.

Put the second super on top. If all the frames contain drawn comb, this works well. If you have only a few drawn combs, distribute them as I advised earlier. You could take one or two partly drawn combs from the bottom super and put them into the second one. The gaps thus created in the first must be filled with the frames displaced from the second super. If you leave them empty, the bees will build free-hanging combs in the space which can cause a few problems later.

The same procedure will apply with the third super. When the third super is full of bees, one or both of the two below should be ready to remove. It may well be, however, that only some of the combs in each will be ready. What is ready? When nectar in cells has been ripened into honey, the bees will cap over the cell with light/white wax. To remove honey and extract it without any danger of fermentation, at least two-thirds to three-quarters of the surface of a comb should be capped over. Combs like this in a number of supers can be concentrated together into one box and taken away.

Removing supers

Put the supers to be removed to one side and replace the remainder in their original order. If you have been wise enough to budget for and acquire a fourth super and frames, add this on the top. Next, insert an 'escape board', making sure you have it the correct way up so that the bees go down into the body of the hive rather than up into the supers you want to clear of bees. Place the supers to be removed on top. The escape board can have one or two Porter bee escapes fitted. The Porter bee escapes act like one-way valves. Bees confined in the original first and second supers want to get out and the only way is through the escapes. They enter through a hole in the upper surface and then push past two pairs of springs into the box below. One end of each spring is secured to the middle of the escape and the other end is free. The two springs in each pair are adjusted so that there is just a bee space gap between the free ends. Bees cannot find their way back through this gap and the supers are cleared of bees. There are other patterns known as Canadian escapes. These follow the same principle but have no moving parts. They can work well and can clear the bees out very rapidly. This is the type I tend to use.

Adding a fourth super allows you to remove honey without causing congestion in the colony. You need to remove the honey when it is ready because of granulation. Honey is made up from many sugars all of which are present in nectar. The two main sugars are fructose (fruit sugar) and glucose. The least soluble of

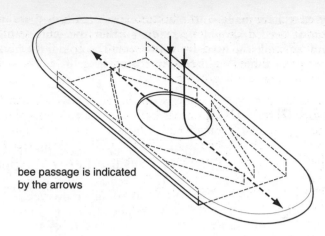

bee passage is indicated
by the arrows

figure 22 a Porter bee escape

these is glucose. Fructose is always the most plentiful but when
the proportion of glucose is high, it will come out of the solution
in the form of sugar crystals or 'granulate'. Honey from oilseed
rape is high in glucose and will granulate in the comb if it stays
too long on the hive or gets cold. So, you must remove the
honey and extract it, probably at least twice, say in mid-May
and at the end of June. The later flows, such as those from ling
heather, are over in most areas by the end of July and the honey
can be removed then.

Returning supers to the hive

After extraction, the sticky (wet) supers can be put back on the
hive. The bees will get quite excited by this, so do it at dusk after
they have finished flying for the day. Proceed as follows:

1 Take off the roof.
2 Place it upside down by the hive.
3 Put the wet super(s) on it.
4 Remove the inner cover and put it on top of the wet super(s).
5 Pick up the super(s) and inner cover all together and place the
   whole lot on the hive.
6 Replace the roof and walk away.

This takes one or two minutes. The bees will repair any damage
to the comb and, because the super now contains drawn comb,
it will be filled faster.

You need to have made your plans to extract the honey, and the equipment needed should be ready, so that the work can be undertaken while the honey is still warm. The colder the honey gets, the more difficult it is to spin out of the cells.

Your honey crop

If the honey is still liquid, the comb can be cut out of the frames and broken up. Using a suitable straining cloth, the broken comb can be drained like fruit in a jelly bag. In fact, you could use a jelly bag if you have one. The comb can even be cut into suitably sized pieces and eaten as 'cut comb'. However, if you want to do this, you need to plan ahead and put unwired foundation in your super frames! It may sound topsy-turvy, but you can stop or delay granulation of oilseed rape honey by putting it in your deep freeze. Many people are put off eating cut comb by the beeswax, but believe me, if you cut a piece of comb from the frame and spread it on bread (or toast) and butter, it is as if the wax vanishes. Whatever happens, the result tastes wonderful – unless you have false teeth!

However, this method may only work for small amounts or in your first season, unless you keep colony numbers down to one or two so that you may be able to consume all of your honey harvest. Even keeping colony numbers low may not be sufficient. In the 1950s, the *British Bee Journal* reported that a South African beekeeper harvested over 360 kg from a single colony! It is also known that an Australian beekeeper averaged over 320 kg of honey per colony from 400 colonies. The 'official' English record is for over 180 kg harvested from a colony in 1940.

Honey extractors

Even if these figures seem wildly improbable to you, the average crop from a National hive in the United Kingdom is 10–14 kg. Sooner or later you will need a honey extractor, which comes in two main types – tangential and radial.

The tangential type has a revolving cage, the mesh of which supports the entire comb face. As the cage spins, centrifugal force tries to make the honey in the cells both on the inside and outside face of the comb fly to the walls of the barrel. This has several consequences.

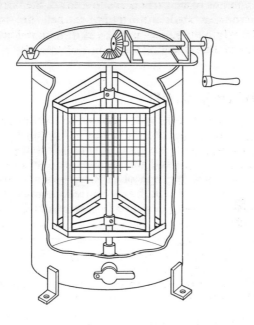

figure 23 a tangential honey extractor

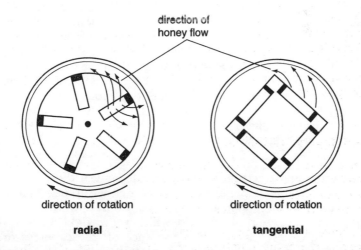

direction of
honey flow

direction of rotation

direction of rotation

radial

tangential

figure 24 the mode of extraction for radial and tangential extractors

- The comb has to be turned, end over end, to partially extract honey from each side in turn until the cells are empty.
- The self-spacing frames, such as Hoffman and Manley, hold the face of the comb away from the internal cage and the comb face is unsupported.
- Enthusiastic rotation can break combs completely out of the frames.

The radial pattern holds the frames in a circular cage, top bars to the outside, bottom bars to the central shaft, like the spokes of a wheel. Both sides are extracted at once. Provided the comb is attached to the frame and the frame is held straight, breakages are few. As combs become older, they become stronger. New combs can be very soft and fragile so treat them gently. Hoffman and Manley frames are designed for use in a radial extractor.

Extraction

Appliance manufacturers and their agents carry all the equipment you need. However, to start with, extraction can be achieved with very simple equipment.

Uncapping

For this, you need a sharp knife or a knife with a serrated edge. I suggest you choose one with a long slim blade rather than a broad blade. Most kitchens have a suitable one. It needs to have a blade at least 200 mm long. It is easier to uncap a shallow (super) frame with a blade that is longer than the frame depth. You will also need a clean plastic washing-up bowl and a strip of wood, say 40–50 mm wide, which is long enough to sit across the bowl. Cut two notches on one surface so that these locate over the edges of the bowl and give it stability. Cut a recess in the opposite surface which is wide enough to hold the end of the frame lug (see Figure 25).

Place the support across the bowl and the end of the frame in the recess. You have the choice of cutting either up or down to remove the wax cappings from the cells. I find upwards easiest but a slip can cause problems. Use whichever method you feel is safest and easiest for you. You can slice off the cappings or simply break them open. An uncapping fork is often used to lift off cappings but you can still extract the honey if the cappings are simply broken by scratching them. In this case, the cappings

figure 25 uncapping a frame

will be collected with the extracted honey which is then strained to remove them.

New combs are very soft. If you play the jet from a hot-air paint stripper gently across the surface of the comb, the cappings will melt away almost instantaneously. The honey doesn't get hot and the 'uncapped' comb can be put straight into the extractor. Don't turn the extractor handle too swiftly or run its motor too fast or for too long. The last drops of honey are extracted at the expense of adding fine bubbles of air to your crop.

Assessing the crop

Supers hold various weights of honey. A rule of thumb is to count about 9 kg for every 'full' super. It will give you a rough idea of the containers you will require to hold your honey when you have extracted it.

If the honey is for your own use, it can be run, or drained if you are using a jelly bag, into whatever containers you desire. If you have bought honey buckets or have acquired food-grade polyethylene containers, I would use them and wait. Twenty-four hours later, especially if the honey is kept in a warm place, the wax particles and air foam will have risen to the surface and can be skimmed off. They are perfectly edible, or you can add them to the bees' winter feed later. What is underneath will be pretty clear. If you wish to sell it, then it must be filtered more finely – all appliance manufacturers sell equipment for this purpose.

Honey filters most easily if it is warm and you will find the finest filters clog most rapidly. Solid honey can be liquefied by heating it to 43–9 °C if it contains really stubborn crystals. There is no point going over this temperature. If you want to warm honey that is already liquid so that it filters more rapidly, then 40 °C is sufficient. If you overheat your honey, it must be sold as 'industrial' or 'baker's' honey.

Be warned! Don't leave honey running out of taps – keep your eye on it. Most beekeepers eventually have their own 'honey all over the floor' story. With care, you may be the exception. I knew a beekeeper who took two steps into his extracting room, leaving a slipper behind at each one. He had left the tap open! Honey flows silently!

The extracting room

Where you extract has to be clean and hygienic. Most small-scale beekeepers use the domestic kitchen. You must have water handy for washing hands and equipment and for washing the floor. This must be available from two sources, so put some in a bowl and use the sink for the other. Common sense should prevail.

You need to arrange your equipment so that one task follows easily and is close to the next.

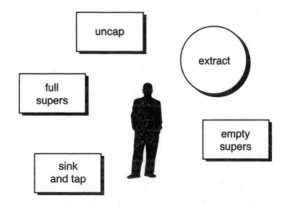

figure 26 layout of the honey extracting area

When you move combs that might drip, try to move them over the equipment. Wash your hands and any handles as they become sticky. If any honey falls on the floor, try to wipe it up using a damp cloth. A wet cloth appears to remove it but the honey simply dries to a bigger sticky patch.

A honey crop approaching a three-figure total may well prompt you to buy a honey tank from which to bottle your crop. Alternatively, you can buy honey taps that can be fixed to larger buckets like those you might buy for home brewing. Check the bee appliance catalogues. They are likely to offer what you want. Occasionally, you can pick up a second-hand extractor at sales or auctions of bee equipment.

Honey can be used in all sorts of ways. I used to know an old beekeeper. He and his wife used 35 kg of honey a year just to sweeten their early morning cups of tea. Don't be overwhelmed by an apparently large crop. It will all go! If you want to sell your honey in England, you must comply with the Honey (England) Regulations 2003. Similar legislation applies in other parts of the United Kingdom. A copy of the Regulations can be found on the Food Standards Agency website or by entering 'Honey Regulations' into an Internet search engine. In the US, the latest situation regarding Honey regulations can be checked on the website of the National Honey Board (www.nhb.org). I recommend that you extract in clean conditions, use standard honey jars and buy your labels from the main appliance manufacturers.

Beeswax

The other thing that bees provide is beeswax. This is produced by worker bees as small flakes from glands on the underside of their abdomens. They hang in clusters and the secreted wax is passed forward to their mandibles where it is manipulated and chewed and used for building combs and cappings. Honey is the fuel for this process. Various estimates have been made for the amount of honey required to make half a kilo of wax. These range from 3.5 kg to 9 kg. Wax production is a natural process and quite normal for bees to undertake. Wax itself, of course, has a value, even if it is only used to exchange for foundation. The cappings and the wax you filter out of the honey you have extracted are part of your harvest. It has been estimated that around 800,000 of these wax scales are required to make half a kilo of wax.

Small amounts of sticky cappings can be put on your hive inner cover, supported on a grid such as a cake cooling rack, making sure that the feed hole is open. You will need to put an empty super around this before replacing the roof so that robber bees cannot gain access. Bees from the colony will then come up and remove any honey. Often, they will also build fantastic shapes from the loose wax.

The cappings can then be washed. If you do this, you must use soft water. Chemicals in hard water will combine with the wax, just like it does with soap, to be lost in a kind of scum.

Cappings wax looks very clean but there will be a dirty residue when it is melted. I suggest that if you have only a very small amount, you save it each year until you have enough and then clean up all your wax cakes in one go. Wax can be discoloured by contact with metals such as iron or zinc when it is molten. Even aluminium gives it a dull colour. Stainless steel and glass are the safest containers to use.

If you want to enjoy your wax, it can be turned into polish (see Chapter 10). It can also be moulded or cast into candles. The best and easiest moulds to use are those made of white silicon rubber. Although they are expensive initially, with care they will last for many castings.

You can melt wax in a microwave, although it won't melt on its own. Add soft water and this will heat the wax and melt it.

In small quantities, wax can be filtered through surgical lint, nappy liners or similar closely woven cloths. Of course, some wax will be lost on the filter so it pays to keep your stash until you have plenty to clean. Wax is shown at honey shows and the criteria used for judging include colour, cleanliness and aroma. Producing a perfect block of wax requires skill and experience. You may well get drawn into the world of showing in the fullness of time. For some people, beekeeping is merely one of the steps on the way to showing.

Storing supers

At the end of the season, when the last honey has been extracted, I think it is best to put the wet supers back on the hive, over the inner cover. Do this in the evening, near dusk. Let the bees into the supers through the feed hole. The excitement this causes is reduced by the night time. Give the bees access to the supers for several days – maybe a week. By that time, every bit of honey will have been licked up.

The problem of how to store the combs over winter comes next. Beeswax is eaten by the larvae of two moths, the greater wax moth (*Galleria mellonella*) (see Plate 21) and the lesser wax moth (*Achroia grisella*). The greater wax moth chews grooves into the wooden hive surfaces when it pupates. It prefers to eat brood comb. The lesser wax moth is more of a pest of stored supers but both species are a menace.

To avoid the damage caused by these two moths, don't store your supers in warm storage areas such as lofts. Consider one of these alternative storage solutions:

- Store supers of empty comb outside in piles on stands. Don't put a floor at the bottom but use a mouse-proof mesh (for example, a queen excluder) on the top and bottom of the pile and top with a good roof.
- After you have prepared your hives for winter, store them temporarily on top of the inner cover with the feed hole open, topped with a sound roof.
- Store in a deep freeze.
- Place each super of comb in a bin bag, close up and store in a cold place. It is important to protect stored supers from mice.

07 winter

In the United Kingdom, the main testing period for colony survival is the winter. It is this season that intensifies the effect of colony pests and diseases as well as testing the bees' ability to conserve food, stay confined, possibly for long periods, and then develop in the spring. There are many reasons why colonies die in the winter. I believe the main one is starvation, followed by queen failure (loss). Disease comes third. In a perfect world, starvation should never happen. Queen failure or mating with too few drones seem to be more common, at least in my bees.

Bees themselves start winter preparations quite early. Gradually, as the brood nest shrinks after mid-summer, bees tend to store more and more food close to it. Under natural conditions, they begin to form a loose cluster at about 14°C. A thicker shell of bees insulates the centre. If brood is present, they maintain the temperature at 35°C. If there is no brood, the temperature could be 21°C.

As the ambient temperature falls, the insulating shell will become thicker and the diameter of the cluster smaller. One of the critical temperatures involved is that at which a bee can just stay alive and move slowly. This is about 6°C. Clustering bees squeeze tightly together, their hairy bodies keeping in warmth. The fuel used to generate this heat is honey. During the real 'winter', a colony may only consume 200–300 g of honey per week so the main bulk of stores is used to fuel colony development and expansion in the spring.

Generally speaking, dark strains of bees are more thrifty with winter stores than the yellow strains, but a total weight of winter stores of 18 kg should be adequate for most colonies. Colonies housed on double brood chambers or large single ones such as the 14 × 12 might well have collected most or all of the winter stores they require. This is probably the most cogent argument for using double brood chambers. The milder autumns of recent years have enabled bees to collect a great deal of nectar from later-flowering sources such as ivy and water balsam. I have even heard of supers returned to the hive for cleaning being filled up again. However, you can't rely on this happening every year.

Feeding

The safe solution is to feed your bees. First assess what is already in the colony. One British Standard brood comb full of honey will hold about 2.0 kg, that is about 1.15 kg per side.

Don't forget that each comb has two sides and both sides have to be full to produce the 2.3 kg total. Look through the brood nest and assess the amount of stored honey. Roughly estimate the area of food as a proportion of the comb face and equate this to an appropriate proportion of 1.15 kg. After this examination, 'heft' the hive. Hefting means raising the back and side of the hive in turn. Don't lift it high, just enough to slip an imaginary postcard between the hive and the stand. With a little experience of equating the weight you feel to your estimate of honey stores, you will be able to dispense with the comb-by-comb assessment and just heft the hive. A commercial beekeeper once told me, 'I feed 15 kg whether they need it or not'.

Ordinary sugar is the most easily obtainable product with which to feed your bees. The much quoted proportions are a sugar to water ratio of 2:1. I have always thought it simpler to use the dry weight of sugar when deciding what to feed your bees so, in fact, the 15 kg fed by the commercial beekeeper mixed with 8.5 litres of water would probably be close to the 18 kg required. It may well be that the colony already has 15 kg of stores so 15 kg of sugar made into syrup is more than adequate. Recently, supplies of ready-made syrup or fondant using inverted sugar have come onto the market. I think that it would be well worth considering using this system for feeding colonies if you own more than one or two.

Many people will say that the Modified National hive cannot hold all the food the colony requires. My bees don't know that and have been quite happy for the past forty years. Very prolific bees are usually 'greedy' and a British Standard hive is then unlikely to hold sufficient.

Feeders

Feeders come in many forms. The simplest is something like an empty jar or tin, say an old 1-kg jam jar. Pierce the centre area of the lid with about a dozen small holes. Fill the jar with syrup and invert it over a sink or bucket. Very little liquid will pour out and the flow will stop almost instantly. Having proved that, go to your hive with a couple of thin pieces of wood about the thickness of the bee space. Place these across the feed hole, invert the jar (again over a bucket to prevent spillage outside the hive) and place it upside down on the sticks. Surround the jar with a frameless bee box (brood or super), replace the roof on the hive and retire. You should give the first feed at dusk after bees have stopped flying for the day. Bees can get very 'excited' and those finding the feeder will be indicating to other foragers that there is food in the immediate neighbourhood of the hive. If you give this first feed during the day, the foragers will go out looking for it and could start a bout of robbing within the apiary. However, given this message at night, they will just look around inside the hive and find the feeder. Once they know where it is and the first excitement has died down, the jar can be topped up and further feed can be given at any time of day. A strong colony will empty a 1-kg jar overnight. If you have more than one colony, it is prudent to start feeding them all at the same time.

Start feeding with this type of feeder in August and carry on until the second or third week in September. Keep hefting your hive to check the weight. Your bees may be just taking down your syrup or they may actually be harvesting nectar from late flowers. Go by the weight. For your own education, check the frames when you think you have fed sufficient. If you find a large area of unsealed stores, feed a little more. Some strains of bees will have converted much of the food into brood. Keep feeding these a while longer but not later than the end of September or thereabouts. Whatever else you do, make sure your bees have enough. It is best to restrict access of small colonies to the combs that they are covering. Feed them until their combs are full. To make a small colony work too hard in the autumn, collecting and storing winter food, will wear out the winter bees and you need them for colony development in the spring.

There are different patterns of feeders. The smallest, like the jam jar, require you to replenish the food regularly. Bees take down the syrup and store it like nectar. A strong colony with empty combs can empty a feeder very rapidly.

You can buy or make bucket feeders operating on the same principle as the jam jar (see Plate 22). The largest available from the equipment suppliers holds about 5 litres. However, the larger these inverted (or contact) feeders are, the more the syrup will run out when they are turned upside down. In the apiary, the safest plan is to invert the feeder over a bucket. What runs out can be saved for the next feed.

You can also buy round feeders, Miller feeders or Ashforth feeders. With a round feeder, bees enter from below to reach the syrup. If you use a round feeder, you will need to put an empty

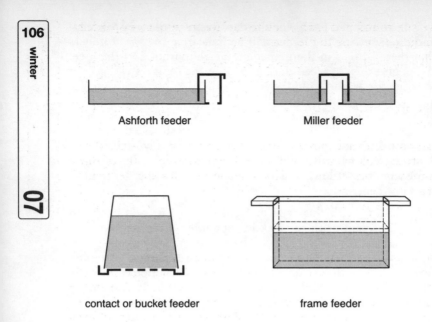

Ashforth feeder

Miller feeder

contact or bucket feeder

frame feeder

figure 27 types of feeder

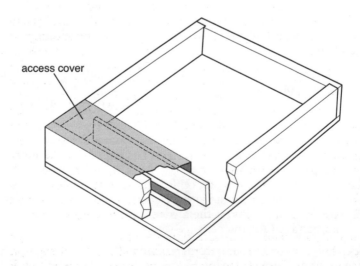

access cover

figure 28 an Ashforth feeder

hive box round it to enable you to replace the roof to stop other bees gaining access to the hive.

Miller and Ashforth feeders use the same principle but cover the whole of the top of the hive. The roof can be replaced directly over them as their outer dimensions are the same as your hive. They are usually made of wood and painted inside to seal the joints.

Such wooden feeders are usually made to hold approximately 10 litres. One filling might be enough to feed your colony completely (see Plate 23). Home-made versions can be made deeper to hold, say, 15 litres. Remember, 4.5 litres (1 gallon) of syrup made from 1 kg of sugar to half a litre of water will contain 3.5 kg of sugar; 15 litres of this strength syrup will therefore be giving the bees over 10 kg of stores.

If you buy a new wooden feeder, it will be watertight. Old ones may leak as the wood shrinks and warps a little. Repaint the insides before you fill and wet them, just to be safe. A slowly leaking wooden feeder certainly encourages robbing. I have known wasps to chew dimples in the wood in order to get at the oozing syrup.

Feeding candy

If you suddenly acquire bees that have no or little food, or a colony that was not fed for whatever reason, it is possible to feed them during the winter. You can buy invert sugar candy or even blocks of fondant icing. If you purchase a large block of this, the simplest way to cut off a hefty slice is to use a wet knife and keep wetting it as you cut. Whatever form it comes in, place the slice directly over the bees – on the frame tops. Take off the inner cover and place the candy or fondant directly over the cluster. Then cover it with something like a plastic food tray followed by some insulation such as newspaper. Surround this with an empty super or an eke so that you can replace the inner cover and roof to keep the hive secure. An eke is made from four pieces of wood of equal depth, nailed together at the corners to make a kind of collar the same dimensions as your hive. It will substitute for a super if you don't have a spare, empty one.

To make your own candy, put half a litre of water in a large saucepan. Start to heat it and gradually add 2.5 kg of sugar, a little at a time. Keep stirring until the mixture starts to boil with a vigorous or rolling boil. Boil like this for three minutes.

Take the saucepan off the heat and stand it in cold water. Continue to stir while the mixture cools, trying to incorporate the cool syrup at the side of the saucepan into the hot mixture in the middle. As soon as the mixture starts to look cloudy, pour it into moulds or into a large metal tray. It will then set. Cut the big block into smaller pieces while it is still warm, just as you would if you were making giant fudge. Keep each piece in a plastic bag until you need it in order to keep it moist. When you want to use it, don't take it out of the plastic bag, just cut a large piece of the bag away and put the bared face of the candy over the bees. There is no need to do this after the bees have started to fly regularly, say in mid-February. You can switch over to syrup at that stage, using a contact feeder. Any unused candy blocks can still be used – diluted into syrup.

Robbing

There are several ways to help your bees against robbers (see also Chapter 8). Large entrances invite invaders. It is sensible, therefore, to restrict the height and length of your entrance. If the floor of your hive has a full-depth entrance, use an entrance block that fills the space, leaving an orifice of around 70–100 mm × 8 mm. A 30 × 8 mm gap will be plenty for a small colony. Make sure there is no other way into your hive apart from that entrance. An added bonus is that an entrance 8 mm high is small enough to keep mice from getting into your colony. You should put the block with the small entrance in place soon after you have removed the supers.

Don't leave honey, wet frames or supers where bees can find them. Don't spill any syrup feed on the outside of the hive or the ground. This can start robbing. It is much easier to prevent robbing from starting than it is to stop it.

Ventilation

Bees positively ventilate their hives during the active season but this may not be possible while they are clustering in the winter. In areas of wet winter weather, you may have to take greater precautions to help your bees. Roofs should not leak and hives should be sheltered from the wind, particularly if it carries rain. They should be off the ground on stands about 300 mm high so that air may circulate underneath. You may need to weight the roof to prevent it blowing away.

My own belief is that however damp the climate, it is more important to provide the means for damp air to escape from the hive than to be too concerned about the size of the entrance in your attempts to keep the inside of the hive dry.

What I suggest is this. Firstly, remove any covering you have placed over the feed hole in the inner cover. Secondly, use small strips of wood or nails to raise the inner cover by 2–3 mm. This allows air to exhaust gently into the space just beneath the roof. The gap is too narrow to allow bees to exit and, of course, too small to allow wasps or robber bees to enter.

If you create this gap too soon, your bees will block it with propolis. Create the gap by mid- to late-October. When you have done so, put on the roof and leave your bees alone except for an occasional visit. With experience, if you find that the combs grow a bloom of mould in the winter, try removing the two outside ones to leave a gap at either side of the brood box. Of course, they should be replaced as soon as bees start to become active on a regular basis around February. Please relate all the dates I give to your own area. What is right for Cornwall will be too early for Staffordshire or Aberdeen.

08

pests and diseases

In this chapter you will learn:
- about pests – human and animal
- what brood should look like
- about brood problems
- about insect pests
- about parasites
- about robbing and drifting.

Bees have been kept by humans for a very long time and a great deal is known about the pests and diseases that threaten them. Recently, because we humans move things around so much, we have actually added some 'foreign' complaints to the list.

Pests

The following deals with the most common pests. There are others but there is not space to cover them here.

Mice

These creatures get into hives and equipment. They chew wooden equipment and ruin and devalue it. They eat stores, especially pollen, chew holes in the combs and drag in grass and leaves to make their nests. Occupied hives suffer in the winter. At any time bees are active, their stings will put mice off completely. However, in winter, the bees will not leave the cluster and mice can gain access.

I lump all small rodents under the title 'mouse'. I am told, and I believe, that mice can squeeze through a gap 12 mm high. Mice cannot squeeze through a 12 mm hole so a bee-space-high entrance will stop them.

Mice cannot chew through metal, even a soft metal like zinc, so if you store your queen excluders on top of the inner cover, even if the roof fits badly, this should stop rodent access that way. A shallow 8-mm floor or a similar height entrance in an entrance block is just as effective. If you want to have a deeper entrance during the winter, then you must protect it with a mouseguard. One form is a strip of zinc, with 9.5-mm holes punched in it, which is pinned over the full entrance. Bees can easily get through the holes, as can air, but mice cannot. Mouseguards should replace entrance blocks by late October or early November. Rats are kept out of the hive by the same means that you use to exclude mice. Rats are bigger so they do more damage. If they get into stored equipment, there is a further problem. They tend to dribble urine as they move about and can pass on nasty complaints such as Weil's disease. It is advisable to replace your combs if this happens and give other equipment a good scrubbing with products that kill all known germs.

Wasps

Wasps can be a severe trial for bees. Each year as I watch wasps trying to gain access to beehives, I am reminded that this contest has been going on for the past 30 or 40 million years. Bees can cope, but with a little help, they can cope more easily.

There tend to be years when wasps are few in number and others when they are a real nuisance. There is a tendency, but not a rule, that 'wasp' years alternate with quieter ones. Since you do not know quite when they are going to occur, always assume the worst.

The best way to help your bees against wasps is to make the hive bee-tight (and hence wasp-tight). All hive roofs, for example, should fit closely. All ventilation holes or slots should be covered with metal gauze. Since wasps are much the same size as bees, the entrances to colonies should be so small that, at active times, bees have to queue a little to get inside. Such entrances are much easier for the colony to defend against invaders. Lots of beekeepers insist that large entrances are mandatory but I have known a colony fill four supers with honey using an entrance measuring 8 × 12 mm. Wasp problems are at their peak in August and September. They can continue through October and into November and in one year I saw wasps still trying to rob bees in December.

Wasps will try to get in through any gap. They will fly away if challenged by bees. The attempts by wasps to gatecrash a hive stimulate bees to guard the entrance. I know of no way to stop wasps coming to try their luck, but strong colonies with a small entrance will be able to keep most of them out.

Wasps do all of this because they are hungry. The adult wasp eats sugar in one form or another. Adults will work flowers to collect nectar for their own benefit. Wasp grubs are fed on protein, often in the form of insect bodies that have been masticated to a liquid by the adult wasp. The grub then rewards the adult with some sweet saliva. In the autumn, the wasp colony starts to 'wind down', having produced fertile queens and drones for next year's nests. The food supply available from the grubs becomes restricted so the adults turn to stealing it from us – and our bees.

Small bee colonies are the most vulnerable. Their entrances should be restricted to one bee space, (8 × 8 mm). I think it also helps the colony if you move the brood nest so that it is close to the entrance. This brings the guards closer to the point of danger. Except in warm weather, even really large colonies of bees will not adequately defend an entrance that stretches for the full width of the hive and is the full depth of a deep floor. Bees are most likely to defend the middle of the gap rather than the ends, allowing wasps to slip in at the sides. In cold-way hives, the brood is usually in the middle, hence my advice about reducing the entrance and moving the brood nest. It is only common sense. However, some colonies are too small to help. You, the beekeeper, should assess your colonies and unite the weak to the strong. Remember to kill the queen in the colony you dislike most. Remember also that the weakest colony goes above the sheet of newspaper.

Bees

Bees rob one another, and my experience of bees is that they will all rob, given the opportunity. Bee robbing can be spectacular but again, as with wasps, provided the colony is strong enough and its entrance is small enough to be easily defended, the bouts of robbing will peter out.

Exposed honey combs, sticky equipment, multiple ways into a hive and small colonies are all things to avoid. Weak colonies you wish to keep intact are better moved to a completely different site where there are no other bees. If you move them to another apiary, they seem to carry a message with them and the bees at the new site will start robbing them again. Avoid examining colonies during the day, or stop examining them as soon as bees from other colonies start showing an interest. Remember that honey bees and wasps can wriggle through holes that look unbelievably small. Quiet, silent robbing like this can deprive a small colony of all its stores. It is as well to note that once inside the hive, wasps and robber bees are left well enough alone to fill up their crops. Guards only make a real effort to keep intruders out at the entrance. If this were not so, no wasp could survive inside a crowded hive. I have seen live wasps in glass-sided observation hives remain alive for hours. Help the guard bees do their job.

Birds

Birds eat insects and they will eat bees. Swallows, martins and swifts are all capable of eating bees, but bees can cope. The ability of colonies to make up losses in worker numbers has been effective for many thousands of years. European Bee Eaters (*Merops apiaster*) have been known to nest here. If the species ever became resident, we might have to add its name to our list of pests.

The green woodpecker can present severe problems to bee colonies. In certain areas, these birds appear to learn what beehives are and in the winter they are capable of smashing fist-sized holes through hive walls. I have seen them eating wax and honey. I expect they eat bees too. Bees in hollow trees must be a natural part of the woodpecker winter diet as they have a very great tendency to attack occupied hives in preference to unoccupied stacks of equipment. I expect they hear bees just as they hear the grubs and other food that they seek naturally in tree trunks.

Hives can be protected against woodpeckers by wrapping them with small mesh (25 mm) chicken wire through which bees can still fly into the hive. This requires a strip 4.5 times the length of one side of the roof and 35–40 cm deep. Cover the box to the floor but leave enough protruding above the top of the roof to fold over. The extra length allows a vertical overlap. Leave the mesh in place until March or April. If you remove it carefully and roll it up, it should last many years. A less expensive alternative is to pin heavy duty plastic to the roof. The pieces should hang down to the hive floor. They can be large panels to cover each side completely or more narrow strips pinned on side by side to cover all four sides. Cut off the pieces on the front just above the entrance to enable the flying bees an easier access.

Livestock

All apiary sites that are likely to be invaded by livestock should be fenced. This should be common sense. Don't take the risk. I was once asked by an eminent beekeeper if he could 'get away' with not fencing a site located in a sheep pasture. He should have known better. Always fence stock away from hives. Livestock can knock hives over which is a potential disaster in winter.

Vandals

I have known children break into gardens, turn hives over and throw stones at them. This is much more common in out-apiaries but still happens relatively rarely. All I can suggest is that your hives are kept out of sight and/or painted in subdued colours. It is even possible to buy bee veils in army camouflage pattern.

Wax moths

See 'storing supers' (Chapter 7).

Diseases

Read the following descriptions through carefully. Read the description of healthy bees just as carefully. If you feel there are any diversions from health in your bees, I strongly suggest that you consult with an experienced local beekeeper or an officer from the National Bee Unit (NBU). Some diseases (see table) are notifiable and it is your duty under the law to report them. Response from the NBU is usually very quick. Each US state has its own inspection programme, carried out by regional inspectors. These only inspect colonies for American Foul Brood but will generally give other beekeeping advice. A new beekeeper can learn a great deal from these experienced people.

Bee diseases can be divided into two groups – larval diseases and those affecting adult bees. For the new beekeeper, it may be too much to learn every little nuance of a disease at first. However, all new beekeepers should first learn the signs of health in brood.

Eggs

The egg should be placed in the centre of the cell, curved over slightly and, as your eye travels towards the centre of the brood comb, eggs should be in cells next to those containing larvae (see Plate 24).

Larvae

These should range in size from about egg size to the point when the larva fills the bottom of the cell (see Plate 24). They should be pearly white. Like eggs, they should be in curved patches as part of the normal brood pattern. There should be approximately twice as many larvae as there are eggs.

Sealed brood – pupae

The brood nest should be dominated by the amount of sealed brood – twice as much as the number of larvae. The cappings will be dark brown on old comb, lighter on newer comb and just off-white over larvae in brand new comb. The cappings of worker brood should be slightly domed and smooth (see Plate 25). Some bees show brood cappings with a dimple. The dome on cappings on drone brood will be more pronounced (see Plate 8).

You should see this each time you look at your bees. At least one examination in the spring and one in the autumn should be dedicated to looking at brood alone to check its state of health.

Symptoms of the major brood diseases are set out in the table. The last two are not strictly diseases but are still problems to be identified and dealt with.

Disease	Appearance of brood
American Foul Brood (AFB) Caused by the bacterium *Paenibacillus larvae* subsp. *larvae*	Cappings are sunken and look greasy. They are dark and may be perforated. Larvae are white-to-brown. Will 'rope' if pulled out of the cell with a matchstick. Dry to a hard scale which is difficult to remove. Can smell.
European Foul Brood (EFB) Caused by the bacterium *Melisococcus plutonius*	Larvae are an off-white colour and lie in unnatural positions in their cells. They can have a 'melted' appearance. Some larvae die in sealed cells but most deaths are in unsealed cells.
Sac Brood Caused by a virus	The larva dies after it has been sealed and spun its cocoon. The virus prevents the larva making its final moult. It is rare for many larvae to be involved. There is no known cure. Looks similar to AFB but the dried larvae can be easily removed from their cells.
Chalk Brood (see Plate 26) Caused by the fungus *Ascosphaera apis*	The larvae dies after its cell is sealed. Cappings can show whitish marks. The larva is hard and chalky. Some develop fungal spores and look grey or black.

Laying workers	Multiple eggs laid in the same cell. Worker cells have a domed capping. There is no normal sealed brood and much stored pollen. The colony deteriorates as the proportion of drones increases.
Chilled brood	This is not a disease. It is easy to avoid unless brood is left exposed for several hours. Chilled larvae die and turn black.

Some knowledge of brood diseases is important if only because a suspicious beekeeper can either deal with a problem or have it dealt with quickly. In addition, some bee diseases are notifiable. If you know or suspect these diseases are in your colonies, you are required *by law* to notify the proper authorities. In this case, the proper authority in England and Wales is the National Bee Unit in York. In Scotland, you should contact the Scottish Executive Environment and Rural Affairs Department and in Northern Ireland, it is the Department of Agriculture and Rural Development. Contact details can be found in Taking it further.

At the time of writing, the National Bee Unit employs experienced beekeepers on a regional basis as Bees Officers who inspect colonies for the presence of both European Foul Brood (EFB) (see Plate 27) and American Foul Brood (AFB) (see Plate 28). If your colony is found to have EFB, it may be treatable or it could be destroyed. If it found to have AFB, I am afraid it will be destroyed. Compensation is available through Bee Diseases Insurance Ltd.

The situation regarding notifiable bee diseases is under review. The best way to keep up with the current position is to join your local beekeepers' association to receive the latest information.

In the case of EFB, you will have a choice of treatments depending on the colony size, the level of infection and the time of year. The colony may be treated using the shook-swarm method, using the antibiotic oxytetracycline, or destroyed. These options will be fully explained to you by the Seasonal Bees Officer (SBO) if the unfortunate situation arises.

The bacterium which causes EFB (*Melisococcus plutonius*), does not produce spores but can live on equipment. If your colony becomes infected, after treatment go over the internal surfaces of the hive boxes, the floor and the inner cover with a blowtorch until the wood turns a dark chocolate brown. The method of shaking a colony onto clean foundation can be very effective but a cure is not a foregone conclusion.

Paenibacillus larvae subspecies *larvae*, the causative agent of AFB, is much more resistant. The spores can survive the temperature of molten beeswax and even that of boiling water. They can live for many years, at least 30 or 40, and can quite happily re-infect bees after that time. Before the National Bee Unit was established, AFB levels were at about ten per cent. At the moment, levels are at one per cent of colonies inspected, which is probably an irreducible minimum.

You should welcome the visit of the Seasonal Bees Officer (SBO). For the new beekeeper to see his or her bees handled by an expert is a valuable educational experience. Since AFB levels are kept down to one per cent the visit will almost certainly be positive. If, however, the disease is suspected, the SBO will conduct field tests and, if necessary, the comb will be sent off for laboratory confirmation. One conclusive test is to take a small wooden rod (a matchstick!) and stir it into some larval remains that are light brown in colour. AFB is confirmed if, when the matchstick is pulled from the cell, a string of material comes with it. This string or 'rope' can be pulled out for a few centimetres.

If AFB is confirmed, the bees are destroyed. All the frames, bees and honey are burned in a deep hole and the remains buried. The hive boxes are scorched with a blowtorch which has a temperature sufficient to kill the spores.

Varroa

There are no colonies of bees in the United Kingdom that are not under threat from this parasite. It was first noticed as a parasite of the Eastern honey bee (*Apis cerana*) and named *Varroa jacobsoni*. It probably transferred to our honey bee in several regions such as Siberia, although research has shown that the mite affecting our bees today is a new related species, *Varroa destructor*. It was first found in Devon in 1992. It soon became obvious that it was already well established and had probably been present for at least two years. Speculation as to how it got here is pointless. It is here and we have to deal with it.

Varroa destructor is a mite (see Plate 29). Both the adults and young feed on honey bee blood (haemolymph). The mites survive our winters on adult bees or in brood if there is any present. However, they can only reproduce in brood. Pregnant female mites enter a cell containing a larva just before it is due to be sealed. They feed on the developing pupa and start to lay eggs. The first egg produces a female, the second a male and the rest are female. The nymphs pass through various life stages. The male mates with his sisters and, in worker cells, the eldest pregnant female leaves the cell, tucked between the abdominal plates of the newly emerging bee. The remaining females are too immature and die in the cell, as does the male. In drone cells, which are preferred by the mites, more than one pregnant female offspring mite can be produced which means that when a colony has drone brood, the *Varroa* population can develop very rapidly. A colony of bees can support a very high mite population. This is quite surprising as the mite is 1.6 mm across. It would be like us being parasitized by fleas the size of a dinner plate. There are excellent advisory leaflets on this and other bee health topics available from the National Bee Unit which is part of the Central Science Laboratory (see Taking it further). Do get hold of as many as you can.

The real problem comes from diseases to which the bees are already prone, namely their viruses. These are taken up by the *Varroa* mites as they feed and they get back into bees by the same process, and prove to be more virulent than they were before. Without treatment, an affected colony will last only a few years. Since the advent of *Varroa*, wild colonies have largely disappeared. Those that do occur don't last. To survive, bees need treatment.

It used to be easy. Two products, Apistan® and Bayvarol®, consist of plastic strips impregnated with an appropriate pyrethroid miticide. These strips are hung between the combs in the brood nest for six to eight weeks, generally in August or September. However, *Varroa* has become resistant to these treatments. This means that the mite has to be dealt with in a variety of other ways. This regime has been given the title 'integrated pest management' or IPM. Put simply, this means attacking the mites in various ways to keep the population in the colony as low as possible. Only in this way – low mite numbers – can the beekeeper control the effects of viruses.

Pest management techniques

Drone brood removal/sacrificial drone brood

In the spring and early summer, bees are likely to build drone comb when given the chance. To encourage this, either place one or more shallow frames in the brood box or fit a shallow sheet of worker foundation in a brood frame. In both cases, provided the hive is level, the bees will build natural comb in the space provided. This is very likely to be drone comb at this time of year. As soon as the brood the queen lays in this comb is all sealed, it can be cut off and burnt, thrown away or destroyed. Wild birds like eating drone pupae. *Varroa* feeding on the drone pupae will, of course, die with it. One or two combs like this in the brood nest will help to control mite numbers.

The 'shook-swarm' method

Basically, this is a method where the bees are shaken off their brood combs onto replacement combs of foundation. If two or three old drawn combs are included with these, the queen will lay in them first and, as soon as this brood is all sealed, they can be removed as well. The removed brood is destroyed along with the *Varroa* mites. This procedure can be carried out when the bees attempt to swarm naturally. A queen excluder needs to be placed under the brood box for the first few days after the procedure (see below).

You will need:

- brood box
- set of frames or foundation
- queen excluder

The method is as follows:

1 Find and cage the queen and put her in a safe place.
2 Move her colony to one side.
3 Place a queen excluder on a new floor on the original site and then add the new brood box.
4 Remove two or three combs of foundation to make a gap.
5 Shake all the bees from the original brood box into this gap.
6 Make sure all the bees clinging onto the original box, floor and inner cover are also shaken into the new box.
7 Release the queen into the new brood box.

8 Replace the queen excluder over the brood box and add any supers.

9 Replace the roof.

If you are doing this when the bees are preparing to swarm, there is a danger that the bees will continue their attempts for a while. The excluder below the brood box will stop the queen leaving. As soon as you see the brood nest developing, you can remove the lower excluder. Again, you can include one or two old brood frames in the new box and they, and the trapped *Varroa* mites, can be removed when the brood in them is sealed. An Apiary Guide covering this technique has been produced by the magazine *Bee Craft* as part of a set on Integrated Pest Management (IPM) techniques (see Taking it further).

Other treatments

Apart from Apistan® and Bayvarol®, the only treatment legally permitted in the United Kingdom is Apiguard®. This is a treatment based on thymol. Administered according to instructions, it can be 95 per cent effective. Another treatment available is the hive cleanser Exomite®, a powder of carnauba wax containing a small amount of thymol. The powder is electrostatic and is attracted to the bees' bodies. The main effect here is achieved because the powder causes the mites to loosen their grip on the bees and they fall off. If you are using this treatment, it is essential to use a mesh floor so that the mites fall through, away from the brood nest. Exomite® can be used because it is designated a hive cleanser rather than a treatment. Legal and effective treatments in the US are Apiguard®, ApiLifeVAR and Mite Away II™. Other legal treatments are becoming ineffective as mites become resistant.

The latest distribution maps available (www.nationalbeeunit. com) show where pyrethroid-resistant mites have been found. At the moment, there are still places where they have not been found. However, new beekeepers should accept that distribution will become widespread and assume that all mites are resistant to Apistan® and Bayvarol®. Start your beekeeping career on this basis and employ IPM techniques from the beginning.

Other substances will kill mites. Oxalic acid is much favoured in this regard. However, at the time of writing, it is not a legal treatment. It may well become one. All beekeepers need to keep abreast of the latest developments in honey bee treatments.

Adult bee diseases

Nosema

The primary adult bee disease throughout the world affects the bee gut or stomach. It is caused by an organism known as *Nosema apis* (see Plate 30). Worker bees clean up mess in the hive by licking and biting with their mandibles. In doing this, they accidentally eat *Nosema* spores. These develop in their stomachs and severely interfere with their ability to digest food, particularly pollen. This can cause dysentery, although colonies can suffer from dysentery without having *Nosema*. The other main effect of the disease is to shorten the affected bee's life. As far as the colony is concerned, this affects the way it builds up in the spring. It either builds up slowly, stays the same size, or dwindles. The population can dwindle to the extent that the colony cannot maintain its viability and dies.

Research by Bailey and Ball (see Taking it further) shows that the majority of apparently normal colonies have some level of *Nosema*. An antibiotic called Fumidil B® can kill the active stage in the bee. However, since the disease carries on year after year as bees pick up spores on the combs, you really need to change these as often as possible as well. *Nosema* lives on as spores in bee droppings (faeces) on the comb. As soon as the weather is good enough for bees to fly from the hive regularly to excrete outside, the infection gradually disappears, which is why the colony builds up slowly only reaching full size in late summer.

Treatment by feeding Fumidil B® in the winter syrup in autumn should be combined with shaking the colony onto clean combs, as described above. However, many colonies appear to cope without any extra help. I have not fed Fumidil B® for many years, preferring to unite 'sick' colonies to each other and eventually change the combs of the combined stock. In this way, I consider that I am reducing the number of susceptible colonies and replacing them from the successful ones.

All beekeepers should be aware of and know as much as possible about bee diseases. It never hurts to keep up with the latest information.

Amoeba

This protozoan, *Malpighamoeba mellificae*, affects the equivalent of the bee's kidneys. It is passed on by bees which ingest the resting stages of the organism, similar to the way that Nosema is spread. The clean comb treatment for *Nosema* will definitely help a colony affected by *Amoeba*.

Acarine

This disease, which I suppose should more properly be called an infestation, is caused by a mite (*Acarapis woodi*). It was first identified in the Isle of Wight and is therefore also known as the Isle of Wight disease.

The mite enters the first pair of trachea of young bees. In there, it breeds, eventually causing a complete blockage. The actual effect on the bee itself seems to be a shortened life. The effect of this on the colony is that the colony is unable to build up in the spring. If enough bees are infested, then the colony may die. The earliest cases noted in 1906 involved the mass crawling of bees out of the colony but this rarely seems to happen today.

Eventually, treatments were developed which usually involved a vapour being introduced to the hive and being 'inhaled' by the bees. Health and Safety Regulations now mean that none of these treatments is any longer legal in the UK.

The only legal treatment for Acarine (known as tracheal mites) in the US is menthol.

On the other hand, the disease seems to have lost its virulence and I have only had to treat colonies on two occasions in the past 40 years.

Some imported bees seem to be more susceptible to acarine than most of our indigenous bees. With the bees that many of us keep, it seems simpler to let the susceptible colonies die out and replace them with those that show resistance.

A full diagnosis of acarine is made using a low-power microscope to examine the breathing tubes (trachea) in the thorax. In the past, identification was on the basis that these trachea were coloured (rather than being pale ivory) or black. We now know that normal-looking, pale ivory trachea can in fact be the home for plenty of acarine mites.

By the time you are reading this, there may well be treatments that are not currently available. I can only suggest that you keep an eye on the present situation and keep close contact with other beekeepers and beekeeping associations.

Robbing and drifting

These aren't diseases but the main way that diseases are passed from one colony to another. When bees 'rob out' a dead or weak colony, they are just as likely to pick up American Foul Brood, European Foul Brood, *Nosema* or anything else. It is easier to work hard to stop the advent of robbing than to stop it once it has started.

Drifting is the tendency for the bees from one colony returning from foraging flights to accidentally enter another one. This is a prime way for disease to be passed on. Bees tend to 'drift' along rows, carried by the prevailing wind. Colonies at each end of the row tend to 'get' more honey. They fill more supers and bees seem to be attracted to taller colonies. Bees are also confused by regular patterns in a 'tidy' apiary. Such regular arrangements need landmarks to help bees identify their own hives. These can be permanent features such as plants and shrubs. My choice for an apiary site layout would be a rough circle with hive entrances facing outwards. You, of course, then work the hives from inside the circle, so there should be gaps between the hives. As it is, hives set in pairs on hive stands should be at least 45–60 cm apart.

Africanised honey bees

These are a hybrid between European honey bees and honey bees from South Africa. The African bees were imported into Brazil for experiments to see if honey crops could be improved by controlled cross-mating with European honey bees already there. Some swarms of African bees were accidentally released into the wild. Rather than dying out, these feral colonies have thrived and expanded. They have increased their range up through Mexico and into the Southern United States. They are renowned for their highly defensive nature although beekeepers are learning to manage these colonies, which have to be kept away from centres of human habitation.

Africanised bees are not found in the UK and are unlikely to survive our climate conditions.

09

the beekeeping year

In this chapter you will learn:
- a month-by-month guide to the beekeeping year
- a quick-glance guide to what is happening.

Here we are going to take a quick overview of the year from the point of view of you, the beekeeper. My comments for each month are based on my own experiences in the English Midlands. I hope you will be able to adjust my advice for your own area, using your common sense.

January

Pay occasional visits to your bees simply to check that everything is as it should be. Livestock can break down fences occasionally. Mouseguards can be pulled off or drawing pins can ping out, letting mice in (this doesn't happen with shallow entrances!). Your bees will fly on sunny days when it is warm enough. The bees will use these cleansing flights to defecate. They may even collect pollen from winter-flowering plants such as Christmas rose. Bees will even fly out over snow when the sun is shining (see Plate 31). The reflection of sunlight from the snow seems to confuse them and they fly into the snow, probably because there is more ultra-violet light. They die of cold on the snow. If you live in a district where snow is likely, it is advisable to shade the entrance.

February

In cold seasons, February will be much the same as January as far as bees are concerned. However, inside the cluster, the queen will definitely be starting to lay and brood production will begin. Many colonies have been found to have brood present in winter. Some have it present intermittently, some permanently, while others are brood free. Whatever the situation, the lengthening days stimulate brood production. In winter, bees will fly whenever the temperature permits. This increasing activity coupled with brood rearing will mean the bees start to make inroads into their winter stores. Check the state of the stores by hefting the hive. If you fed enough in the autumn, things will almost certainly be fine. On warm days, bees may well clear out some of their dead sisters. This can be quite normal. It is also normal to find little crumbs of wax on the floor or alighting board. This is a sign that bees are beginning to open the cells and use their stores. Another sign of increased brood production is a melted circle in snow or hoar frost on the hive roof. As time goes on and as ambient temperatures increase, bees will be more and more able to go on useful foraging trips. Their main requirement at the moment is to augment their pollen stores with fresh pollen and they will collect this from plants such as snowdrops, crocuses and *Viburnum tinus* (Laurustinus). They will also be collecting water to dilute their winter stores. Make sure the supply you provide is in a sunny spot, as taking in cold water will chill the bee and could prevent it from being able to fly back to the colony.

March

Colonies should be flying well every day when weather conditions are suitable. More and more plants will be being exploited by your bees and they will be making the greatest inroads into their stores as the rate of brood rearing increases. Pollen loads being brought back to the hive should be large and very noticeable. It might well be worth using just a little smoke to have a quick look inside at the top bars. The bees should look just as strong in late March as they did in September. If they are not, or if there is a lot of excreta visible because they have dysentery, your bees may not make it. Now is the time to remove any props used to lift the inner cover. Cover the feed hole and leave the bees to their own devices. If the hive is very light or you feel that too much food is being used, you may begin to feed your bees. In spring, use dilute syrup – a 1:1 ratio is about right. As a quick fix, just put the sugar in a vessel and double that volume with water.

In fact, spring feeding may be started at any time from January or February onwards if it is obvious that bees are going to be able to take flights on a nominal weekly basis. Feeding dilute stores when bees cannot fly from the hive gives them the problem of getting rid of liquid waste. The inability to do so can cause dysentery.

By the end of the month, it may be possible to look at the brood nest in your colonies. Do not do this too early. Some bees, if examined at an early stage of brood nest inception, will kill their queen. Wait until the flowering currant (*Ribes sanguinium*) is in flower in your district, or even in your own garden, and you should be able to open your bees when the weather is good.

April

Colony development really accelerates this month. It is important to keep an eye on your bees' activities. Add a super over a queen excluder as soon as the brood box is full of bees. Start regular examinations of the brood nests of larger colonies this month, probably in the second or third week. Record what you see. Remember, regular examination of the brood nest is the only reliable way to detect early signs of swarming. Make sure the equipment you need to deal with swarms is ready. Note the weather and how it affects your bees. You will see how their level of activity rises and falls during the day. They are probably responding to changes in temperature as well as the nectar and pollen yields from plants. Remove any old, broodless frames and combs that you wish to change and replace them with frames of foundation. If you do this, you may need to feed to help the bees make wax to draw out the comb.

Seek experienced help if you think that all is not well.

May

Continue adding supers as each new box on the hive fills with bees. Continue regular brood nest inspections. If necessary, list out the steps of the swarm control method you have chosen to use and have this ready and available so that you can consult it if required. Be ready to remove supers or frames full of honey that you think are ready for extraction. Replace what you take away. Avoid crowding colonies. Attend beekeeping meetings to see bees being handled – and ask questions! No question is silly if you don't know the answer and if you don't ask, you won't learn.

June

Continue as in May. Successful swarm control may mean that the need for very regular inspections is coming to an end. May and June are the main months for swarming. There may well be a 'June gap' in your area where there is little available forage. If so, make sure that the bees are not left without adequate food reserves. If you have a newly mated queen following swarm control procedures, try to have her heading your main honey-producing colony. This will reduce the chances of further swarming to a very low level.

July

Lime trees flower early this month, and clover also starts now if you have any near your hives. July is still the main honey month for most beekeepers. Bees want to consolidate stores this month. Assess how much honey you have in your supers. Remove most of your honey crop when it is sealed and bees don't seem to be storing freely. Bees are not so bothered now about overcrowding. They want to store honey around their brood nest, ready for winter. Don't give more supers unless you know your bees need them. Be aware of crops such as borage that could still be yielding. Willowherb can also continue producing nectar into early August.

August

Remove the supers. Take care not to expose frames or honey. Reduce entrances as wasps may start to be a real problem this month. If your bees have no food reserves in the brood nest, feed syrup. Uniting is possible this month, but if bees are defensive, because of robbing, a few bees may be killed. Do not embark on any extensive beekeeping operations unless they are absolutely necessary. This is the month to start treatment against *Varroa destructor* if you are using Apiguard®, Apistan® or Bayvarol®.

September

Feed the bulk of your winter stores this month. It is better to complete feeding by the third week. The aim is to make sure the final brood nest is surrounded by food. In the Midlands, ivy will be starting to yield. Do not rely on winter stores from this source as the weather may mean that it is unavailable. Check hive weights by hefting.

October

Keep your eye on the wasp situation. Do not forget that honey bees rob each other so check for this as well. Make sure all hive parts are bee- and wasp-tight. Keep entrances small until wasps have gone and frosts have killed their nests. In damp locations, a large entrance may be beneficial but protect your hive from mice. Learn from what other local beekeepers do. Remove *Varroa* treatments on time, in accordance with the manufacturer's instructions. Legally, you are required to record the batch numbers of treatments and the dates of insertion and removal.

November

Raise inner covers for winter ventilation. Try larger entrances in warm, damp districts and protect deep entrances with mouseguards. Protect hives against woodpeckers. Leave your bees alone after this – they must be kept quiet and you must not disturb the winter cluster. Protect hives against livestock. In windy places, place a weight on the roof to prevent it blowing away.

December

Visit your bees every two to three weeks to check that all is well. Read bee books. Review the past year. What worked? What didn't? What additional equipment do you need for next year? Make or buy it now. If you can saw straight, you can make a beehive!

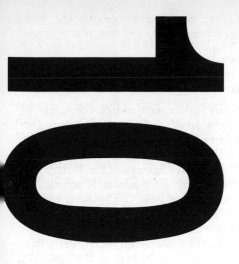

10

hive products

In this chapter you will learn:
- about honey
- about beeswax
- about mead
- about polish
- about candles
- about propolis.

Honey

Honey is the main reason why humans keep bees. In pre-history, the honey was just taken as soon as a bee colony was discovered. Then, by degrees, we came to the present situation where humans give a degree of care in return for the harvest.

What is honey? It is certainly sweet, but unlike the sugar on your table, it has flavours and aromas. Many honeys are very similar in aroma and taste but others are very distinctive. Most honey-producing countries have a honey they consider to be their best. One of the most distinctive honeys of the British Isles comes from ling (*Calluna vulgaris*). The aroma is strong and delightful to those who like heather honey. The deep reddish brown colour is also unique. The honey itself can 'set' like a jelly but becomes liquid when it is stirred. It is, in fact, thixotropic – like non-drip paint. It usually sells for a much higher price than 'ordinary' honey. In reality, the best honey is the one you personally prefer.

Honey is sugar – a natural blend of sugars plus essential oils, traces of minerals, and water. Nectar, which is actually the bees' food, has quite a high water content of 60–80 per cent and this will not keep. Yeast cells enter the nectar and it starts to ferment. In order for nectar to keep, its water content has to be reduced to the point where it is impossible for yeast cells to survive. The bees do this when they convert nectar to honey and most British honeys have a moisture content of around 18 per cent. There is even a law about this. In a document from the FAO and WHO called the Codex Alimentarius, acceptable moisture levels in honey are specified. This means that when you stick a label on your jars declaring it to be honey, you are saying that the contents adhere to the legal description.

A typical honey contains:

Water	17–18%
Fructose	39–42%
Glucose	34–35%
Sucrose	1–2%

Dextrose, proteins, wax, salts, acids (e.g. citrus acid), volatile oils, pollen grains and pigments make up the remainder.

Honey granulates because it is a saturated solution of sugars. In the hive, the honey near the brood nest is kept warm and it remains liquid. However, in the cooler parts of the hive, the least soluble sugar (glucose) forms crystals in the more soluble one (fructose), and the honey granulates. This is a perfectly natural process and granulated honey can be returned to a liquid state by warming to no more than 43–49 °C. However, heating to higher temperatures or for any length of time is not good. This increases the level of a substance called hydroxymethylfurfuraldehyde, or HMF, and you may well then be breaking the rules for honey composition set out in the Codex, which cites maximum permitted HMF levels.

Honey will keep for a long time. I have tasted one that was 43 years old. It was still palatable but had little aroma. I have the gravest doubts, however, that the honey found in recent times in the tombs of the ancient Egyptians was palatable!

If you have comb honey which has been cut straight from the frame, it does need to be eaten up relatively quickly. Fifty years ago, section dishes were common. These were made to hold little square 'sections' of comb. Section frames made from basswood were placed, twenty-eight or twenty-one to a box and placed on the hive like a super. They were fitted with thin, unwired foundation which meant that the resulting honeycomb could be taken out and used directly. Today, comb honey is mostly sold in small tubs similar to margarine containers. The super frame is fitted with unwired, thin foundation and then sections of honeycomb are cut out to fit into the containers. You then just cut off a small piece and spread it on your bread and butter. The wax is so thin and breaks up into such minute pieces that it is hardly noticeable. Many consider that this is *the* way to eat honey and experience the greatest flavour.

Honey can also be added to other foods, for example it can be used to make jams and marmalades. However, whenever you substitute honey for sugar in a recipe you have to allow for the water content in the honey and reduce it elsewhere. You can also divide the normal sugar requirement in recipes and use half honey and half sugar. Honey can be used to sweeten porridge, cornflakes, fruit, tea or coffee. I knew an elderly Russian lady who would use two pounds a week, just to sweeten her tea!

In honey shows, you will find classes for cakes made using honey. If you assume that roughly one fifth, or 20 per cent, of honey is water, then you may be able to adapt an existing recipe to use honey. The following is a recipe used in the National Honey Show in 2004.

Honey Fruit Cake

110 g butter or margarine
175 g honey
2 medium-sized eggs
227 g self-raising flour
pinch of salt
175 g sultanas

Pre-heat the oven to 180 °C, 350 °F or Gas Mark 4.

Cream butter and honey together. Beat eggs well and add alternately with the sifted flour and salt. Beat the mixture well and lightly. A little milk may be added if necessary. Fold in the sultanas and stir well. Prepare an 18-cm tin. Pour in the cake mixture. Bake for 1¼ hours.

Honey can also provide the sugar required to start yeast working for bread making. It can be used as a glaze for roasting, but avoid pouring it onto your meat too soon when cooking as the sugars will burn.

Mead

Mead is diluted fermented honey and is one of the oldest known alcoholic drinks. Mead is described in texts that are thousands of years old and is supposed to pre-date wine made from grapes. Occasionally, you find mead available for sale from a commercial source. For some reason, such mead is almost always made incredibly sweet. I have found that it is only beekeepers who seem to make dry mead. This is a pity because a well-made dry mead can be magnificent, especially if it is drunk chilled like a dry white wine.

Just mixing honey and water will not make such a mead. Although it is acid, a honey/water mixture needs more. The yeast will need more food such as a yeast nutrient. If you are new to wine making, it is much easier to start with a melomel which is a mead made with honey and fruit. Honey is added to fruit juice to bring the sugar content up to the required level. I have found an easy way to make a simple melomel is by using white grape juice purchased from the local supermarket.

Melomel

Each litre of grape juice will need about 100 g of honey to raise the sugar content high enough to make dry mead. You will therefore need about 454 g of honey if you are making 4.5 litres of melomel.

454 g honey
yeast nutrient
1 tsp pectolase (to remove pectin – this is optional)
5 litres grape juice
1 tsp all purpose wine yeast

Dissolve the honey, yeast nutrient and pectolase in four litres of the grape juice. Pour this into a clean demijohn. Add the yeast and fit the air lock containing water. Leave this in a warm place (room temperature is ideal) and the bubbles passing through the air lock will tell you when the yeast is working. At first, this will be at a high rate. You will not have filled the demijohn and the space above the liquid will allow any foaming to occur without an overspill. Keep an eye on the rate of bubbling and when it becomes very slow, use part of the fifth litre of grape juice to top up the demijohn to just below the bung. You can then leave this until the bubbling stops completely and the melomel clears.

Taste your melomel at intervals from the demijohn by taking tiny sips through a drinking straw. If you want a sweeter drink, before the bubbling stops, remove enough melomel to dissolve 100 g of honey and return it to the demijohn. Let this ferment. If you want it sweeter still, repeat the process when the bubbling slows down again.

There are good books on mead making. I heartily recommend *Mead* written by Dr Harry Riches (see Taking it further).

Beeswax

As a new beekeeper, you will not produce much wax at first. Wax is mainly harvested from the cappings removed from honey storage cells during extraction and your first harvest will probably only amount to a few ounces per colony. If your wax has come from cappings it will undoubtedly be sticky with honey. Wash it in soft water – filtered rainwater is fine. Using hard water means you will lose some of the wax as scum.

Before you use your beeswax for polish or candles, you must clean it. One way is to melt the cappings (assuming you have only a small quantity) in a microwave. Put clean, soft water in a glass bowl and add the cappings. Heat in short bursts with the microwave on high. The water needs to be hot enough to melt the wax, at 71 °C, so boiling is unnecessary. If it does boil you might get foaming which will make a real mess if it splashes or overflows the container. You can also melt the wax by putting it in a stainless steel saucepan together with clean soft water. Again, try to avoid vigorous boiling.

Once melted, cover the bowl and its contents with insulating material to slow down the rate of cooling. This allows 'heavy' dirt to sink. When the wax is cold and has set, the dirt can be scraped off the bottom of the block.

This treatment may be sufficient to clean the wax for your purposes. If you want to clean it further, melt it down again, this time in a bowl over hot water. Pour the resultant liquid through a filter. Cut both ends off a clean tin and fasten your filter material over one end with a strong rubber band. You can use pieces of old sheet, nappy liners, etc. to strain out the dirt. Pour the molten wax through the filter and into your mould which can be something like a clean aluminium foil tray used for a take-away meal.

Once you have your block of wax, you can decide what to do with it. If you have managed to plan ahead, you can make your candle or polish while the wax is still liquid. If you want to use it later, you can leave it to set and it will shrink away from the sides as it cools and be easy to remove. You can then wrap it in a plastic bag until you want to use it.

Your first wax harvest will probably be sufficient for you to make some beeswax polish. A simple polish can be made by shredding or breaking the beeswax into small pieces and putting 280–300 g into half a litre of pure gum turpentine. If you can't find this locally, you can obtain it from beekeeping suppliers. Put the mixture into a sealed container and shake it every day or so. The wax will dissolve very slowly without heat.

You can, of course, achieve the same polish using heat. Use a glass or stainless steel bowl over a water bath. Put the turpentine in the bowl and add the shredded beeswax. Put the bowl on the water bath and keep stirring until the wax is dissolved. Pour the mixture out into suitable containers. Traditionally, these are metal polish tins but you can also buy plastic ones. If you want to sell this beeswax polish, in addition to any other labels you may wish to use, you will have to include a suitable hazard warning label because of the turpentine.

This polish can be made more cheaply using white spirit instead of turpentine. However, the result will smell like paint and will not be as pleasant and fragrant as that made using turpentine.

A much lighter, softer polish can be made by incorporating water into the mixture. To mix the oily wax and turpentine with water, you will have to use another substance to help form an emulsion. The easiest to obtain is soap and I suggest that the easiest as far as preparation goes is soap flakes. If you wish to use solid soap, you will need to shred it. The aim is to create a 'hot' solution of wax melted into turpentine and a separate solution of soap dissolved into the water which is at least at the same temperature.

Polish recipe

170 g shredded beeswax
470 cc pure gum turpentine
235 cc soft water/rainwater
15 g soap flakes

First prepare more containers than you think your polish will fill. Heat the shredded wax and turpentine in a glass or stainless steel bowl over a water bath. In a separate container, pour boiling water over the soap and stir to dissolve it. This solution can be hotter than the wax/turpentine solution but not colder.

Remove everything from the heat. Pour the soap solution through a fine strainer into the wax/turpentine solution and stir with a balloon whisk. There may well be undissolved soap particles and you don't want them to spoil your polish. Continue to stir as the mixture cools and the polish will eventually begin to thicken. At this point, pour it into the containers. It is better to have a half-filled container than a congealing mass in the bowl.

For the best results, beeswax polish requires one more ingredient. It is one that only you can add – elbow grease!

Candles

Making candles is one of the many extensions of beekeeping. Until you have sufficient quantities of your own beeswax, you can buy it from other beekeepers to make candles and other products.

Traditionally, church candles have to contain a certain proportion of beeswax. There are symbolic reasons for this, but also beeswax candles make little smell when they burn, compared with the stink from tallow candles, and this must have influenced this tradition as well.

Wicks in candles are designed to curl over when lit so that the tip burns away in the hot edge of the flame. Wick is available in different thicknesses, each capable of completely burning away the wax in candles of different diameters. However, wick designed for use in a 25 mm (one-inch) diameter candle made from paraffin wax will not be sufficient to burn away the same size candle made from beeswax, so when you purchase candle wick, make sure that the supplier confirms that it is suitable for the size of *beeswax* candle that you plan to make.

I suggest you start candle making with a rolled candle. For this you need some one-inch beeswax candle wick and a sheet of unwired British Standard brood foundation. You can, of course, use foundation of any size, but you will have to experiment to determine what size wick you need or ask your supplier for advice.

Cut the wick about 15 mm longer than the short side of the foundation.

If the foundation is cold and stiff, you can warm it sufficiently to make it pliable by passing hot air from a hair dryer over it. Place the foundation on a flat surface and press the wick into the soft wax about 5 mm from one of the short edges. Try to keep it straight. Now carefully bend the free edge back over the wick. Start at one end and work up to the other a little at a time.

You can then re-warm the wax if necessary. Put the foundation back on the flat surface and start to roll the covered wick over and over, taking the foundation with it, like making a Swiss roll. Try to roll it straight. You can go back to the beginning and start again if you go wrong. Stop rolling when you are 2–3 cm from the other edge. Give this edge an extra warm with the hair dryer (but don't melt it). Then finish rolling the candle over it and smooth the edge down with your thumb. Having made the wax at this end pliable, it should stay put. In order to help the candle light easily, you can dip the protruding wick into a little molten beeswax but this is not vital.

You can shape the top of your candle by trimming the sheet of foundation before you start. Put the foundation on a cutting board or similar. Take a straight edge and put one end at the corner of the sheet and the other about 15 mm down from the corner at the other end. Cut along the straight edge with a sharp knife. Then start rolling from the longest edge and you will end up with a candle with a pointed top. There are, of course, numerous variations on how you cut up your sheet and it is great fun to experiment. Keep the trimmings. You can use these to decorate your finished product.

Rolled candles usually burn well and for a surprisingly long time. Coloured foundation especially for candle making is available from suppliers. There are even competitive classes in honey shows for rolled, dipped, poured and moulded candles so, if you get hooked, there is a whole new world out there waiting to be conquered!

Other uses of beeswax

Did you know that in the past beeswax was used on sweets like wine gums to keep them separate? Unfortunately, because beeswax is generally contaminated today with chemicals used for *Varroa* treatment, it can no longer be used and it has been replaced by carnauba wax which comes from the leaves of the carnauba tree (*Copernicia prunifera*), a member of palm family.

Beeswax has many other uses including cold cream, lipstick, encaustic art, modelling, thread waxing, lost wax casting, waxing pins used in lace making, polishing army boots and making mouthpieces for didgeridoos!

You can save up your wax. With care, you will probably be able to harvest 500–1000 g of wax for every 45 kg of honey you extract. Eventually, you will have enough to consider part-exchanging it for foundation. Foundation for brood frames is generally sold in packs of ten sheets. Several beekeeping equipment suppliers will exchange your clean wax for ready-made foundation. You will probably have to give them a greater weight of wax than you receive in foundation but you will have a product that is ready to use and of the correct dimensions. You can make your own foundation, but this is a much easier way to obtain it.

Propolis

Bees collect this natural resin from tree buds such as poplar and take it back to their nest in their pollen baskets (see Plate 32). Shreds of propolis are pulled off the buds and packed using the middle pair of legs with the result looking like a shiny pollen load.

In the nest, propolis is used to fill up cracks and crevices in the walls. I have seen the interior of a rotten log held firmly in place by propolis collected by its honey bee inhabitants. Similarly, I saw the same thing inside a huge cavity in a church wall which left the stonework in a better state than the eighteenth-century bodgers had!

Bees are sensitive to the texture of propolis and they will collect similar substances that have the right degree of stickiness. I have read unconfirmed reports of them collecting warm tar from roads and semi-dry paint from a tractor firm in Australia. In the latter case, the damage was so serious that all the newly painted tractors had to be resprayed. Because of this, I think it is probably just a coincidence that propolis has such advantageous abilities to kill bacteria and fungi and therefore helps to counteract these organisms in the hive.

The Greek word *propolis* means 'before the city'. Many races of bees use it to control the size of the colony entrance, often reducing this to a few bee-sized holes in a curtain of propolis which otherwise covers the entire entrance. These small entrances can be more easily guarded.

The antiseptic properties of propolis are presumably needed by the tree to help protect its buds from attack. The properties give propolis applications in pharmacology, although there is not a large market in the United Kingdom. There are a few people who are allergic to propolis and it can cause severe eczema.

Propolis was a constituent of the varnish Stradivarius used to coat his famous violins.

Propolis is also used by bees to varnish the inside of the hive and to strengthen beeswax comb. It is usually brown or yellowish brown in colour, and occasionally reddish brown. One of the beekeeper's problems is getting this sticky stuff off fingers, gloves and garments. You can remove it from your hands with methylated spirits which loosens and dilutes the propolis when you rub it in. Quickly, before the mixture has had a chance to dry, wash your hands thoroughly. You can also remove propolis from your hands with a well-known proprietary kitchen and bathroom cleansing cream – and plenty of rubbing.

Stain removers are available for the removal of a number of different of different stains, and those designed for use on resin and tar will remove propolis from clothing. It is stubborn stuff, however, and you will have to take care not to leave a solvent mark. Washing in a hot solution of washing soda and detergent will remove it from suitable fabrics.

Items such as queen excluders clean up better if you leave them for a while because the stickiness of propolis reduces over time. Better still is to leave them outside until you get a cold day when you will find the propolis will scrape off easily. However, when you clean a queen excluder, make sure it is fully supported on a flat surface before scraping it with the flat end of a hive tool. If you catch the edge of the tool in one of the slots you can bend the metal and unwittingly enlarge the hole so that, when you use it the next season, the queen can nip through and start laying eggs in the supers!

If you want to specifically collect propolis in quantity, you can buy special screens which go over the brood box and the bees pack propolis into the holes. The screens are then put in the deep freeze and the propolis is supposed to flake off when the whole thing is frozen.

taking it further

There are many useful books on bees, beekeeping and associated subjects. I have tried to list a few in each area but for a wider selection, contact the bee booksellers and appliance manufacturers. Some of the books in the following list are out of print but it is worth trying to obtain them. Your local beekeeping association may well have copies in its library or you may be able to borrow them from your public library.

Further reading

Andrews, S. W. (1982) *All About Mead*, Mytholmroyd, Northern Bee Books

Apimondia Standing Commission on Beekeeping Technology and Equipment (1978) *Propolis: scientific data and suggestions concerning its composition, properties and possible use in therapeutics*, Bucharest, Apimondia Publishing House

Aston, David and Bucknall, Sally (2004) *Plants and Honey Bees: Their Relationships*, Mytholmroyd, Northern Bee Books

Bailey, L. and Ball, B. V. (1991) *Honey Bee Pathology*, 2nd ed., London, Academic Press

Bee Craft Ltd (2005) *The Bee Craft Apiary Guides to Bee Diseases*, Stoneleigh, Bee Craft Ltd

Bee Craft Ltd (2005) *The Bee Craft Apiary Guides to Integrated Pest Management*, Stoneleigh, Bee Craft Ltd

Brown, R. H. (1989) *Beeswax*, 2nd ed., Burrowbridge, Bee Books New and Old

Davis, Celia F. (2004) *The Honey Bee Inside Out*, Stoneleigh, Bee Craft Ltd

Food Standards Agency, website: http://www.food.gov.uk/foodindustry/regulation/foodlawguidebranch/foodlawguidech03/foodlawhoney

Free, John B. (1987) *Pheromones of Social Bees*, London, Chapman and Hall

Hansen, Henrik (nd) *Honey Bee Brood Diseases*, Copenhagen, L Launs

Hodges, Dorothy (1984) *The Pollen Loads of the Honeybee: A Guide to their Identification by Colour and Form*, Cardiff, International Bee Research Association

Hooper, Ted (1998) Guide to Bees and Honey, 3rd rev. ed., Totnes, Marston House Publishers

Jeffree, E. P. (nd) 'The size of honey-bee colonies throughout the year and the best size to winter.' In: *Honeybee Biology*, compiled by J. B. Free, Central Association of Bee-Keepers

Kirk, William (1994) *A Colour Guide to the Pollen Loads of the Honey Bee*, Cardiff, International Bee Research Association

Lindauer, Martin (1978) *Communication among Social Bees* (Harvard Books in Biology, No 2) 2nd ed., Cambridge, MA, Harvard University Press

Mace, Herbert (nd) *Bee Matters and Beemasters*, Harlow, The Beekeeping Annual Office

Morse, Roger A. and Flottum, Kim (eds) (1997) *Honey Bee Pests, Predators and Diseases*, 3rd ed., Medina, OH, The AI Root Co Ltd

Munn, Pamela (ed) (1998) *Beeswax and Propolis for Pleasure and Profit*, Cardiff, International Bee Research Association

Riches, Harry (2003) *Insect Bites and Stings: a Guide to Prevention and Treatment*, Cardiff, International Bee Research Association

Riches, Harry (1997) *Mead: Making, Exhibiting and Judging*, Burrowbridge, Bee Books New and Old

Seeley, Thomas D. (1995) *The Wisdom of the Hive: The Social Physiology of Honey Bee Colonies*, Cambridge, MA, Harvard University Press

Various, *Beekeeping in a Nutshell*, Mytholmroyd, Northern Bee Books

Von Frisch, Karl (1967) *The Dance Language and Orientation of Bees*, Cambridge, MA, Harvard University Press

Waring, Adrian (2004) *Better Beginnings for Beekeepers*, 2nd ed., Doncaster, BIBBA

Wedmore, E. B. (1979) *A Manual of Beekeeping*, 2nd rev. ed., Burrowbridge, Bee Books New and Old

White, Joyce and Rogers, Valerie (2000) *Honey in the Kitchen*, rev. ed., Charlestown, Bee Books New and Old

White, Joyce and Rogers, Valerie (2001) *More Honey in the Kitchen*, rev. ed., Charlestown, Bee Books New and Old

Winston, Mark L. (1987) *The Biology of the Honey Bee*, Cambridge, MA, Harvard University Press

Useful contacts

Bee book suppliers

B&K Books, Newport Street, Hay-on-Wye, Hereford HR3 5BG

Bee Books New and Old, The Weaven, Little Dewchurch, Hereford HR2 6PP (www.honeyshop.co.uk)

Northern Bee Books, Scout Bottom Farm, Mytholmroyd, Hebden Bridge, West Yorkshire HX7 5JS

Bee disease insurance

Colonies can be insured against losses resulting from European Foul Brood and American Foul Brood with Bee Diseases Insurance Ltd, c/o Llys Gwyn, Cefn Mawr, Newtown SY16 3LB.

Beekeeping associations

England: British Beekeepers' Association, National Beekeeping Centre, Stoneleigh-Park, Kenilworth, Warwickshire CV8 2LG (www.britishbeekeepers.com)

Scotland: The Scottish Beekeepers' Association. 20 Lennox Row, Edinburgh EH5 3JW (www.scottishbeekeepers.org.uk)

Northern Ireland: The Ulster Beekeepers' Association. 26 Coach Road, Comber, Newtownards, Co Down BT23 5QX (www.ubka.org)

Southern Ireland: The Federation of Irish Beekeepers' Associations. Ballinakill, Enfield, Co Meath, Ireland (www.irishbeekeeping.ie)

Wales: The Welsh Beekeepers' Association. Graig Fawr Lodge, Caerphilly CF83 1NF

United States: Each state has its own beekeeping association.

Beekeeping magazines

American Bee Journal, published monthly by 51 S. 2nd Street, Hamilton, IL 62341, USA (www.dadant.com)

An Beachaire, published by the Federation of Irish Beekeepers' Associations, Scart, Kildorrery, Co Cork, Ireland

BBKA News, published every two months by the British Beekeepers' Association, National Beekeeping Centre, Stoneleigh Park, Warwickshire CV8 2LZ, (www.britishbeekeepers.com) (available only as part of Association Membership)

Bee Craft, published monthly by Bee Craft Ltd, 107 Church Street, Werrington, Peterborough PE4 6QF (www.bee-craft.com)

Bee Culture, published monthly by The AI Root Co, 623 W. Liberty Street, Medina, OH 44256, USA (www.BeeCulture.com)

Beekeepers Quarterly, published quarterly by Northern Bee Books, Scout Bottom Farm, Mytholmroyd, Hebden Bridge, West Yorkshire HX7 5JS (www.beedata.com)

Beekeeping, published ten times a year by Devon Beekeepers' Association, Leat Orchard, Grange Road, Buckfast, Devon TQ11 0EH

Gwenynwyr Cymru, published quarterly by the Welsh Beekeepers' Association, Golygfan, Llangynin, St Clears, Carmarthen SA33 4JZ

The Scottish Beekeeper, published by the Scottish Beekeepers' Association, Milton House, Main Street, Scotlandwell, Kinross KY13 9JA (available only as part of Association membership)

Most local beekeeping associations also publish their own magazine or newsletter.

Beekeeping equipment manufacturers and suppliers

Agri-Nova Technology Ltd, The Old Forge, Wendens Ambo, Saffron Walden, Essex CB11 4JL (www.agri-nova.biz)

BB Wear, 1 Glyn Way, Threemilestone, Truro, Cornwall TR3 6DT (www.bbwear.co.uk)

Bee Equipped, Brunswood Farm, Brunswood Lane, Bradley, Ashbourne DE6 1ON

Betterbee Inc, 8 Meader Road, Greenwich, NY 12834 (www.betterbee.com)

BJ Engineering, Unit 2, Worcester Enterprise Centre, Shrub Hill Trading Estate, Shrub Hill Road, Worcester WR4 9FG

BJ Sherriff International, South Cornwall Honey Farm, Carclew, Mylor, Falmouth, Cornwall TR11 5UN (www.beesuits.com)

Brunel Microscopes Ltd, Unit 2, Vincients Road, Bumper's Industrial Estate, Chippenham SN14 6NQ (www.brunelmicroscopes.co.uk)

Brushy Mountain Bee Farm Inc, 610 Bethany Church Road, Moravian Falls, NC 28654 (www.brushymountainbee farm.com)

B Weaver Apiaries, 16481 County Road 319, Navasota, TX 77868 (www.beeweaver.com)

C Wynne Jones, Ty Brith, Pentre Celyn, Ruthin LL15 2SR

Circomb, 29 Glamis Road, Dundee DD2 1TS

Compak, Unit 1B, Electra Park, Electric Avenue, Witton, Birmingham B6 7EB

Dadant and Sons, 51 South 2nd Street, Hamilton, IL 62341 (www.dadant.com)

EH Thorne (Beehives) Ltd, Beehive Works, Wragby, Market Rasen LN8 5LA (www.thorne.co.uk)

Freeman and Harding Ltd, Unit 18, Bilton Road, Erith DA8 2AN

Harrison Smith French Flint Ltd, Rich House, 40 Crimscott Street, London SE1 5TE (www.frenchflint.com)

Honeymaker UK, The Homestead, Hiams Lane, Hartpury, Gloucestershire GL19 3DQ

KBS, Brede Valley Bee Farm, Cottage Lane, Westfield, Hastings TN35 4RT

Maisemore Apiaries, Old Road, Maisemore, Gloucester GL2 8HT (www.honey-online.co.uk)

Mann Lake Ltd, 501 1st Street South, Hackensack MN 56452-2589 (www.mannlakeltd.com)

National Bee Supplies, Merrivale Road, Exeter Road Industrial Estate, Okehampton, Devon EX20 1UD (www.beekeeping. co.uk)

Park Beekeeping Supplies, 17 Blackheath Business Centre, 78B Blackheath Hill, London SE10 8BA (www.pbs.kentbee. com)

Stamfordham Ltd, Heugh House, Heugh, Newcastle-upon-Tyne NE18 0NH (www.stamfordham.biz)

The R Weaver Apiaries Inc, 16495 CR 319, Navasota, Texas 77868 (www.rweaver.com)

Walker T Kelley Co Inc, PO Box 240, 807 West Main Street, Clarkson, Kentucky 42726-0240 (www.kelleybees.com)

Western Bee Supplies Inc, PO Box 190, Polson, MT 59860 (www.westernbee.com)

Vita (Europe) Ltd, 21/23 Wote Street, Basingstoke RG21 7NE (www.vita-europe.com)

Government agencies

England and Wales: Central Science Laboratory, The National Bee Unit, Room O2G11, Sand Hutton, York YO41 1LZ Tel: 01904 462559; website: www.nationalbeeunit.com

Scotland: The Scottish Executive Environment and Rural Affairs Department. Tel: 0131 244 3377

Northern Ireland: The Department of Agriculture and Rural Development Northern Ireland. Tel: 028 9052 4426

Southern Ireland: Teagasc, Bee Diagnostic Unit, Oak Park Research Centre, Carlow, Co Carlow, Ireland

United States: There are five bee research laboratories which are part of the US Department of Agriculture (www.usda.gov). The USDA also controls the National Honey Board (www.nhb.org)

glossary

Abdomen The third and largest body segment of the bee which contains the heart, stomach and intestines. In the worker, it contains the sting and wax glands. In the drone it contains the testes. In the queen it contains the ovaries and the spermatheca.

Acarine Disease caused by a mite (*Acarapis woodi*), which infests the bee's tracheae leading from the first pair of spiracles on the thorax. Known as the Isle of Wight disease because if was first noted there in 1906.

Alighting board A strip of wood, usually fixed to the hive stand or as part of the floor, which protrudes in front of the entrance, giving a platform on which bees can land before running into the hive.

American Foul Brood Caused by a spore-forming bacterium (*Paenibacillus larvae* subspecies *larvae*). Develops in the gut of the larva and kills it after the cell is sealed. This causes sunken and perforated cappings which indicate the problem. It is a notifiable disease.

Amoeba An protozoan, *Malpighamaeba mellificae*, which affects the bee's equivalent of the kidneys.

Anaphylactic shock A severe reaction resulting from an acute allergy to bee venom. It may cause sudden death unless immediate medical attention is received.

Antenna One of a pair of 'feelers' on the head of the bee which carries sensory cells for touch, smell and vibration. Plural: antennae.

Apiary The place where one or more hives are kept. The largest I have seen held 140 colonies.

Apiculture The practice of keeping bees.

Apiguard® A thymol-based treatment for control of *Varroa destructor*.

Apistan® A slow-release polymer strip pyrethroid formulation specifically designed for use in beehives for control of *Varroa destructor*.

Ashforth feeder A wooden feeder the same size as the hive box. There is one syrup reservoir to which bees gain access from one side.

Bait hive A hive placed to attract stray swarms.

Bayvarol® A slow-release polymer strip pyrethroid formulation specifically designed for use in beehives for control of *Varroa destructor*.

Bee hive A container for housing honey bees. It consists of a floor, brood box, one or more supers, an inner cover and a roof.

Bee space The space left by bees between the comb and other surfaces in the hive. It is large enough for queen, worker and drone to pass through.

Bee suit A long, full-body suit, usually made from light cotton and polyester, worn by a beekeeper for protection and comfort while opening beehives or collecting swarms.

Bee tight The situation where there is no way in or out for bees or wasps.

Beeswax A hydrocarbon produced from glands on the underside of the abdomen of the worker bee. Used by bees for comb building and capping cells.

Brace comb Bridges of wax built between adjacent surfaces in the hive.

Brood The immature live stage of the bee. Cells containing eggs and larvae are known as open brood. Sealed cells in which the larvae pupate into adult bees are known as sealed brood.

Brood box The area in which the queen is confined and the brood is reared. One or more brood boxes may be used.

Brood pattern The pattern of concentric swirls of brood at different stages of development. A good brood pattern has few empty cells and indicates that the queen's brood is largely healthy.

Burr comb Wax built on a comb or upon a wooden part in a hive but not connected to any other part.

Canadian escape A board for clearing bees from honey supers. It has no moving parts. Bees pass through narrow gaps and fail to find a way back. They come in many different patterns.

Capped brood See 'Sealed brood'.

Cappings The beeswax covering over a cell. Cappings over honey consist of wax only. Those over brood include hair and other materials.

Cast The same as a swarm in all respects except that it may contain one or more unmated queens.

Caste A different form of the same sex. Bees have two castes in the female form – queen and worker.

Castellations A thin piece of metal into which slots are cut to take frame lugs. They are fastened to the inside upper end of the hive body from which the frames are suspended. They are designed to maintain a constant spacing between frames and are available with differing numbers of slots.

Cells Small, six-sided hexagonal wax compartments making up honey comb. Used to store honey and pollen and to rear the juvenile life stages of bees.

Chalk Brood Caused by a fungus (*Ascosphaera apis*), which affects sealed brood.

Clearer board An inner cover designed to accommodate one or two Porter bee escapes.

Cleansing flight One made by bees which have been confined to the hive for long periods such as in winter or during bad weather. Bees avoid defecating inside the hive and make a cleansing flight when the weather improves.

Cold way Frames arranged at right angles to the entrance of a hive. See also 'Warm way'.

Colony The viable living unit for honey bees comprising a queen and workers. During the summer, male drones are also present.

Comb The mass of six-sided beeswax cells built by honey bees in which brood is reared and honey and pollen are stored. The comb is built in two layers, with the cells in each layer pointing in opposite directions and the layers joined at the base of the cells.

Compound eyes Typical of insect vision. In the bee, the two eye patches are composed of thousands of separate eyes which combine results to give the bee a picture of its world.

Contact feeder One which gives bees direct contact with the contents. It does not cover the whole of the hive surface area and must be surrounded by an eke or empty super so that the roof can be replaced tightly.

Cover board See 'Inner cover'.

Crystallization See 'Granulation'.

Cut comb Natural comb or comb built on thin foundation cut to a size to fit a container for sale.

Cut comb foundation Very thin sheets of beeswax foundation. It is as close as practicable to the thickness of the mid-rib found in naturally built comb.

Dextrose One of the two principal sugars that constitute honey. Also known as glucose. The other principal honey is fructose (levulose).

Drawn comb Foundation where the cells have been drawn out by the worker bees into full depth cells.

Drifting The tendency of bees from one colony to accidentally enter another when returning from foraging flights.

Drone The male bee, reputed to be lazy, but actually works hard at his true function – to fly to drone assembly areas and mate with a virgin queen.

Drone comb Sections of the comb built for raising drones. The cells are slightly larger than worker cells and have a convex, domed capping when sealed.

Drone congregation area The places where drones congregate to mate with virgin queens who travel to the same areas.

Drone layer A queen that lays only unfertilized eggs which develop into drones. This may be because she failed to mate or because she did not mate with a sufficient number of drones and has used up all the sperm stored in her spermatheca.

Dummy frame A slab of wood cut to the same size as a frame to take its place in a hive.

Dysentery Caused by an excessive amount of water in a bee's body. Afflicted bees cannot hold waste products in their bodies and defecate inside the hive. Usually caused by prolonged confinement during winter and early spring and consumption of food with high water content. Often, but not invariably, associated with *Nosema*.

Egg The first stage of honey bee metamorphosis. Eggs laid by the queen appear as small, thin, rods, usually placed in the bottom of the cell. They look like tiny white bananas, about 1.6 mm long.

Eke Four pieces of wood nailed together into a square the same size as the hive. Used to extend the hive when required.

Entrance The elongated space across the front of a beehive through which bees exit and enter the hive.

Entrance block A removable block of wood used to reduce the width of the hive entrance.

European Foul Brood Caused by a bacterium (*Melissococcus plutonius*). Infects the gut of the developing larva and competes for food. Does not kill all affected larvae. A notifiable disease not confined to Europe.

Exoskeleton The hard outside covering of all insect bodies, including bees.

Fermentation The chemical breakdown of honey, caused by sugar-tolerant yeast and associated with honey having a high moisture content. Used to advantage when making mead.

Fertile queen A queen, inseminated instrumentally or mated with a drone, that can lay fertilized eggs.

Flying bees Worker bees old enough to have largely completed their duties in the hive that go out foraging for nectar and pollen. Foraging generally starts at three weeks of age.

Following The annoying habit of some bees to follow and possibly sting another animals coming near to their nest. Some can follow for a good distance from the colony.

Foragers See 'Flying bees'.

Foraging The act of seeking for and collecting nectar, pollen, water and propolis.

Foundation Beeswax sheets impressed with the shape of cell bases and the bases of the cell walls. It can be obtained in sizes suitable for worker and drone cells. It can be strengthened with wires or used without.

Frame Wooden or plastic structure designed to hold bee comb and enable the beekeeper to inspect and utilize it fully.

Frame runner A narrow piece of folded metal fastened to the inside upper end of the hive body from which the frames are suspended.

Frame spacers Plastic or metal spacers which fit over frame lugs and butt up to the spacer on the adjacent frame to ensure constant spacing. Can be narrow or wide. Wide spacing is used only in supers where deeper honey storage cells are desired.

Fructose The predominant simple sugar found in honey, also known as levulose.

Glucose See 'Dextrose'.

Granulation When crystals are formed naturally in honey by the least soluble sugar (dextrose), especially when its temperature falls.

Guard bees Bees that wait at the hive entrance to guard it from invaders, such as foreign bees, wasps, animals or humans. Guard bees give off an alarm pheromone (scent) if the hive is disturbed or threatened, and are the first to fly at and attack the invader.

Hefting The act of just lifting a hive from its support to ascertain its weight.

Hive A man-made or man-modified structure intended as a home for bees. The best hives allow beekeepers to inspect all aspects of bee life.

Hive stand A structure that supports the hive and raises it off the ground.

Hive tool The composite lever/scraper used in the manipulation of a colony.

Hoffman frame A type of self-spacing frame.

Honey The concentrated form of nectar which will keep for a long time. The colour and flavour of honey depends on the flowers from which the nectar is gathered.

Honey crop An organ in the bee's abdomen used for carrying nectar, honey or water.

Honey extractor Machine that allows honey to be extracted from combs so that they can be reused.

Honey flow A heightened influx of nectar into the hive brought about by favourable weather conditions and the availability of suitable flowers.

Honey ripener A holding tank for honey that allows air to rise to the surface before bottling using the tap at the base.

Honeydew The product of sap-sucking bugs such as aphids. Collected by bees when it is diluted by dew. Honeydew flows therefore occur early in the day.

House bee A young worker that stays in the hive and performs tasks such as feeding young larvae, cleaning cells, and receiving and storing nectar and pollen from foragers.

Inner cover A board which is placed over the frames just beneath the roof.

Integrated Pest Management The use of substances and specific manipulations to a colony to reduce the population of *Varroa destructor*.

Invert sugar syrup A liquid sugar syrup formed by inversion, or chemical breakdown, of sucrose resulting in an equal mixture of glucose (dextrose) and fructose (levulose).

IPM See 'Integrated Pest Management'.

June gap A period during June when availability of forage is seriously reduced. This could cause colonies to starve unless the beekeeper checks their stores and feeds if necessary.

Larva The second stage of bee metamorphosis. The grub-like larva hatches from the egg. It then develops into a pupa and changes into an adult.

Laying worker A worker that lays unfertilized but fertile eggs, producing only drones. This occurs if a colony becomes queenless and is not able to raise a new queen.

Levulose See 'Fructose'.

Mating flight The flight taken by a virgin queen when she mates in the air with several drones.

Mead An alcoholic wine-like drink made from honey and water.

Melomel A type of mead made using fruit juice and honey.

Metal ends See 'Frame spacers'.

Metal runner See 'Frame runner'.

Migratory beekeeping The moving of colonies of bees from one locality to another during a single season to take advantage of two or more honey flows.

Miller feeder A wooden feeder the same size as the hive box. There are two syrup reservoirs to which bees gain access from a central slot.

Modified National The commonest hive in use in the United Kingdom. It is a single-walled hive.

Moult The shedding of its skin by a larva to make room for new growth.

Mouseguard A metal strip or similar containing holes that allow bees in and out of the hive but prevent mice from gaining access.

Nectar The sugary secretion of plants produced to attract insects for the purpose of pollination.

Nectar guides Marks on flowers believed to direct insects to nectar sources. They may be visible to the human eye or may reflect ultra-violet and hence be visible only to bees.

Nosema Caused by a microsporidian parasite (*Nosema apis*). This infects the gut of the bee and shortens its life by preventing it from properly digesting its food.

Nucleus hive A small colony, usually on three, four or five frames. Used primarily for starting new colonies, or for rearing or storing queens. Also known as a 'nuc'.

Nucleus A small hive designed to contain three, four or five frames only.

Nurse bees Young worker bees, three to ten days old, which feed and take care of developing brood.

Orientation flight A short flight taken by a young worker in front of or near the hive prior to when it starts foraging in order to establish the position of the hive.

Out-apiary Apiaries established away from the beekeeper's home.

Petiole The narrow waist between the thorax and abdomen of the bee.

Pheromone A substance produced by one living thing that affects the behaviour of other members of the same species. Pheromones produced by the queen help the colony to function properly.

Plastic ends See 'Frame spacer'.

Play cells See 'Queen cell cups'.

Play flight See 'Orientation flight'.

Pollen The part of the plant carrying the male contribution to the production of future generations.

Pollen basket A segment on the hind pair of legs in a worker bee specifically designed for carrying pollen. Also used to bring propolis back to the hive.

Pollen load The pellets of pollen carried by a foraging worker bee in the pollen baskets (corbiculae) on its hind pair of legs.

Pollination The transfer of pollen from the anthers to the stigma of flowers.

Porter bee escape A device invented by an American of the same name. Two spring valves allow bees to pass through one way but not return. Used for clearing bees from supers.

Prime swarm The first swarm to leave the colony, usually containing the old queen.

Proboscis The mouthparts of the bee that form the sucking tube or tongue. Used for sucking up liquid food (nectar or sugar syrup) or water.

Propolis A resinous material collected by bees from the opening buds of various trees, such as poplars.

Pupa The third stage in the development of the honey bee during which the organs of the larva are replaced by those that will be used by an adult. Takes place in a sealed cell.

Queen One of the two variants or castes of the female in bees. Larger and longer than the worker bee.

Queen cell cups The base of a queen cell into which the queen will lay an egg designed to develop into a new queen.

Queen cell An elongated brood cell hanging vertically on the face of the comb in which a queen is reared.

Queen excluder A device with slots or spaced wires which allows workers to pass through but prevents the passage of queens and drones.

Queen substance Complex pheromones produced by the queen. Transmitted throughout the colony through the exchange of food between workers to alert other workers of the queen's presence. Its presence stops worker bees rearing more queens and/or inhibits them from laying eggs.

Queenless The situation when a colony has no queen. If bees have access to worker eggs or very young larvae, they are able to rear a replacement queen.

Queenright The situation when a colony has a living, laying queen.

Quilt See 'Inner cover'.

Retinue Worker bees who attend the queen and care for her needs within the hive.

Ripe queen cell A queen cell that is near to hatching. Bees remove wax from the tip, exposing the brown parchment-like cocoon.

Robbing When wasps, or bees from other colonies, try to steal honey from a hive.

Royal jelly A highly nutritious glandular secretion of young bees, used to feed the queen, young brood and larvae being reared as new queens.

Sacbrood A virus disease which prevents the final larval moult. The larva dies in its larval skin which is easily removed from the cell.

Scout bees Worker bees that search for new sources of nectar, pollen, water and propolis. If a colony is preparing to swarm, scout bees will search for a suitable location for the colony's new home.

Sealed brood The pupal stage in a bee's development during which it changes into an adult.

Sections Honey comb built into special bass wood frames. Generally sold complete. Also available in circular plastic form.

Self-spacing frame A frame in which the upper part of the side bar is extended to touch that of the adjacent frame. Designed to maintain a constant distance between adjacent frames.

SHB See 'Small hive beetle'.

Shook swarm A swarm-sized mass of bees shaken, together with their queen, from one hive into another. Used to control swarming or diseases such as *Varroa destructor* or European Foul Brood.

Skep A beehive constructed from straw which does not contain moveable frames. No longer in general use as a permanent home for a colony. Now often used for collecting swarms.

Small hive beetle A small beetle (*Aethina tumida*) about one-third the size of a worker bee. Dark red, brown or black. Has distinctive fringed antennae. Both larvae and adults eat honey and pollen. Will spoil honey in the comb. It is a growing pest in the US although not yet thought to be present in the UK. It is however a notifiable disease in the UK.

Smoke The product of burning suitable materials. The best smoke for working with bees comes from organic materials such as rotten wood, shavings, dried grass, etc.

Smoker Device that delivers smoke in a precise manner. It was devised by Quinby and improved by Bingham both of whom were American beekeepers.

Spermatheca A special organ in the queen's abdomen in which she stores sperm received from drones during mating.

Spiracles Aperture found on the sides of the thorax and abdomen which lead to the breathing tubes or tracheae.

Sting The defensive mechanism at the end of the abdomen used by worker bees to deter predators. The queen will use her sting to kill rival queens, usually when several are hatching or due to hatch during the swarming process.

Stone Brood A fungal disease similar to Chalk Brood but caused by a different organism.

Stores The weight of honey collected by bees, especially the reserves needed for winter.

Super The box(es) placed on top of the brood chamber to increase the space available for the colony for honey storage.

Sugar syrup A solution of sugar and water used to feed bees.

Swarm A mass of bees not in a hive. The bees may be wanting to establish a new colony or be absconding from a bad environment. It should contain a mated queen.

Swarm cell Queen cells, often but not always found on the bottom of the combs before swarming.

Swarm control Methods used by beekeepers to stop a swarm from leaving the hive.

Swarm prevention Methods used by beekeepers to prevent the physical conditions arising which stimulate a colony to prepare to swarm.

Thin foundation A sheet of foundation which is thinner than that used for brood rearing. Used for comb honey production.

Thorax The second and central part of the bee's body. It contains the flight muscles and has the legs and wings attached.

Tropilaelaps *Tropilaelaps clareae* and *Tropilaelaps koenigerum* are serious parasitic mites which affect both developing brood and adult honey bees, much as the varroa mite does. Its natural host is the giant Asian honey bee (*Apis* dorsalis) but it can readily infest colonies of the western honey bee, *Apis* melifera. Tropilaelaps is not currently known to be in either the UK or the US but is a statutory notifiable pest of honey bees and any suspected infestation should be reported to the appropriate government department. An advisory leaflet on Tropilaelaps is available from the National Bee Unit (www.national beeunit.com).

Uncapping knife A knife used to remove the cappings from combs of sealed honey prior to extraction.

Uniting The act of combining two or more colonies to form a larger colony. Colonies are usually united if one is weak or has lost its queen. It is unwise to unite a small sick colony to a large healthy one.

Varroa destructor A mite that breeds in sealed brood cells, feeding on the larval blood. If it does not kill the developing larva, it can lead to serious deformations such as shrivelled wings. It is the latest pest to affect bees in the United Kingdom.

Veil The see-through but bee-proof garment worn by beekeepers to protect against stings.

Venom Poison secreted by special glands attached to the bee's sting.

Venom allergy A condition in which a person, when stung, may experience a variety of symptoms ranging from a mild rash or itchiness to anaphylactic shock. A person who is stung and experiences abnormal symptoms should consult a doctor before working with bees again.

Virgin queen A young, unmated queen.

Waggle dance The most common communication dance used by bees to indicate a food source over 100 metres from the hive.

Warm way Frames arranged parallel to the entrance of a hive. See also 'Cold way'.

Wax glands Eight pairs of glands on the underside of the last four visible abdominal segments of the worker bee which secrete small particles of beeswax.

Wax moth (greater) The greater wax moth (*Galleria mellonella*) is the most serious and destructive insect pest of unprotected honey bee comb in warmer regions. Wax moths primarily infest stored equipment but will invade colonies where the worker bee population has been weakened. The larvae chew into woodwork to make depressions in which to pupate.

Wax moth (lesser) The lesser wax moth (*Achroia grisella*) has the same type of scavenging habits as the greater wax moth but causes less damage. The adults are similar to clothes moths and are characterized by a yellow head.

WBC hive A double-walled hive designed by William Broughton Carr. Often used in gardens because of its pleasing appearance.

Windbreak A barrier to break the force of the wind blowing onto hives in an apiary. The best windbreak is a thick hedge.

Winter cluster The roughly spherical mass adopted by bees as a means to survive the winter.

Worker The commonest bee in the colony. In a strong colony, there may be 40,000–50,000 workers. They are specialized to undertake the tasks required for the continuation of the colony such as feeding young larvae and foraging for nectar and pollen.

Worker comb Sections of the comb built for raising worker bees. When sealed, the cappings are flat. Also used for storing honey and pollen.

index